METALLURGY
fundamentals

by

Daniel A. Brandt
Mechanical Engineering Department
Milwaukee School of Engineering

South Holland, Illinois
THE GOODHEART-WILLCOX COMPANY, INC.
Publishers

Library of Congress Cataloging in Publication Data

Brandt, Daniel A.
  Metallurgy fundamentals.

  Includes index.
  1. Metallurgy.   I. Title.
TN665.B74 1992 669 91-22280
ISBN 0-87006-922-5

INTRODUCTION

METALLURGY FUNDAMENTALS provides instruction and information on the basic principles of metallurgy. It explores the reasons for the strange ways that metals behave when they are heated and cooled. It explains why the strength and hardness of metal changes after being heated and cooled in certain ways. A knowledge of these principles and characteristics is invaluable to any person who plans to deal with metals as a future vocation.

This text emphasizes the practical aspects of metallurgy. Numerous drawings, photographs, and diagrams show metals in action. Technical descriptions provide information as to how hot to heat metal, how fast to cool it, which metal to use for a given application, and how to measure the effects of heating and cooling on metal.

METALLURGY FUNDAMENTALS clarifies what many industrial processes are all about, so that you can confidently discuss the processing of metals with others in the field. You will understand what is meant by quenching, annealing, normalizing, case hardening, tempering, and crystallization. You will learn to recognize the internal structures of metal.

This text speaks to the reader in down-to-earth language, rather than to the scientist in highly theoretical terms. In many cases: photographs are used instead of lengthy word descriptions; practical examples are given instead of abstract theories; diagrams are provided to show the effects of carbon content in steel and the effects of time and temperature on the heat-treating process.

METALLURGY FUNDAMENTALS is written for those who want to learn the "basics," for those who want to explore the behavior of metals, and for those who want a broad knowledge of the entire field of metallurgy.

DANIEL A. BRANDT

CONTENTS

PRACTICAL APPLICATIONS OF METALLURGY

After studying this chapter, you will be able to:
☐ Define metallurgy.
☐ Explain what a metallurgist does.
☐ Relate how a knowledge of metallurgy can be used to solve industrial problems.
☐ Tell why the study of metallurgy can be a valuable asset.

METALLURGY AND METALLURGISTS

The dictionary defines METALLURGY as "the science that explains methods of refining and extracting metals from their ores and preparing them."

Today, the subject of metallurgy digs deeper into the heart of metals than that. It is more than examining the refinement and extraction of metals from their ores. METALLURGY is the science that explains the properties, behavior, and internal structure of metals. Metallurgy also teaches us what to do to metals to get the best use out of them.

The study of metallurgy actually explores what makes metals behave the way they do. The exploring is done by METALLURGISTS, who are scientists in metallurgy that probe deeply inside the internal structure of metal to learn what it looks like. They seek to understand why the metal changes its structure as it is heated and cooled under many different conditions.

Metallurgy involves all metals, Fig. 1-1. However, this study of metallurgy will deal mainly with iron and steel. Steel is made primarily from iron. Other alloys are added to the iron like a cook would mix in salt, pepper, and onion slices in preparing a mouth-watering stew. See Fig. 1-2.

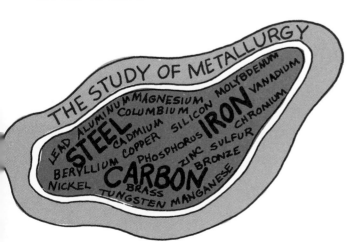

Fig. 1-1. Metallurgy involves all metals, especially iron and steel.

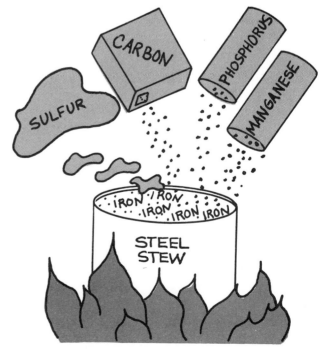

Fig. 1-2. Many alloys are added to iron to produce steel.

There are two reasons why this study of metallurgy is primarily concerned with iron and steel:

1. Steel is the most widely used metal in modern industry.
2. The internal behavior of steel can be predicted.

Forecasting the internal actions of steel during heating, quenching, annealing, tempering, and other heat-treating processes is an exciting challenge. Not only because of the interesting changes that steel goes through, but also because you — as a metallurgist — can predict what the steel will do when it is heat treated.

The examination and knowledge of this predictable behavior of iron and steel is the major thrust of this book on metallurgy.

PRACTICAL EXAMPLES OF METALLURGY IN MODERN INDUSTRY

The way in which metallurgists work is shown in the following examples of how a knowledge of metallurgy is used to solve specific industrial problems.

1. A gear in a machine ran continuously and turned rapidly. It encountered large forces. As a result, the gear teeth showed signs of having worn rapidly. See Fig. 1-3. If a hard, strong material were used to make the gear, it would resist this type of wear. However, most hard and strong materials are also brittle and crack under repeated shock forces.

   Solution: To solve this problem, a metallurgical process known as "case hardening" was used. CASE HARDENING produces a hard surface on the metal part while the interior core remains relatively soft and ductile (workable, not brittle). See Chapter 15 for full details on case hardening.
2. In a particular manufacturing operation, five irregular slots were cut into a large, thin disc. See Fig. 1-4. The slots had to be machined to dimensional tolerances closer than $\pm$ .001

Fig. 1-3. Excessive wear of gear teeth can occur if proper metallurgical processes are not used.
(The Falk Corporation, subsidiary of Sundstrand Corporation.)

(one thousandth) in. ($\pm$ 0.025 mm). After these slots were cut, the disc was installed in a business machine and adjusted until it ran perfectly.

However, problems developed after these machines were shipped to customers. The disc had twisted and distorted after being used only two months. The machining of the slots had created internal stresses in the disc. While the disc was in use, these stresses

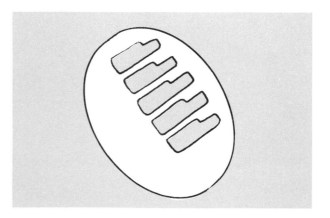

Fig. 1-4. Internal stress caused distortion of this disc until process annealing was used.

gradually relaxed and caused the disc to twist slightly. This distortion caused friction between the disc and another part, which caused the machine to malfunction and bind up.

Solution: This problem was solved by a metallurgical process known as PROCESS ANNEALING. For details, see Chapter 12. Process annealing is a heat-treating process that causes the metal to relax and get rid of internal stresses. In this "disc" application, process annealing caused the disc to distort *before* leaving the factory. Following this relaxing action, a light machining cut was taken to eliminate the few thousandths of an inch distortion. Then, the machine was shipped, free of internal stresses that could have caused distortion at the customer's site if the process annealing had not been done.

3. The cutting tool shown in Fig. 1-5 must be very, very hard. If it is hard — and properly ground and sharpened — it will cut metal cleanly and accurately. However, after a period of use, this cutting tool did not remain sharp. It wore away excessively fast. Then, it did not cut well.

Fig. 1-6. When "tempering" of a cutting tool takes place during a cutting operation, excessive wear of tool can result.

Fig. 1-5. This cutting tool is used on a lathe.

Solution: Again, a knowledge of metallurgy was used to solve the problem. A metallurgical microscopic examination showed that the cutting tool was going through a process known as "tempering" without the machine operator knowing it. See Fig. 1-6.

TEMPERING is a reheating of metal to slightly soften it. Tempering is usually a helpful metallurgical process performed to make the metal more stress-free, distortion-free, and crack-free. However, in the case of this cutter, the unintentional tempering was destroying it.

4. The cutting blade shown in Fig. 1-7 is as sharp as a razor blade. In addition, it must be very hard and strong in order to cut chemically treated paper in a particular industrial application. The problem occurred when the blade did not make clean, smooth cuts.

Solution: It was discovered that a metallurgical process called WATER QUENCHING was used to harden and strengthen the blade. Water quenching, however, also causes distortion that would prevent the blade from cutting cleanly in this application. To solve this problem, a metallurgical process called AIR QUENCHING was substituted for water quenching. In addition, the metallurgist changed the material to a higher alloy tool steel. With these changes, the newly manufactured blades are now making keen and accurate cuts.

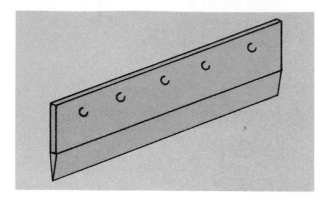

Fig. 1-7. Air quenching technique is used to obtain a hard, sharp cutting blade. Blade will distort (twist slightly out of shape) if water quenching method is used.

WHY STUDY METALLURGY?

Modern industry is dependent on a knowledge of metallurgy. Nearly every kind of manufacturing today is affected by the behavior of metals and alloys. Therefore, anyone who plans a future career in modern industry will find a working knowledge of metallurgical processing to be a valuable asset.

Engineers, technicians, designers, drafters, tool and die makers, and machinists need skills in selecting materials and heat-treating processes. Even production managers and purchasing people can benefit from an understanding of terms such as ductility, hardness, normalizing, and surface hardening. Repair workers, service personnel, and troubleshooters who diagnose causes of equipment failures should be trained to recognize the causes of cracks and excessive wear. They need to know how to examine a material to see whether it has become too hard and too brittle.

Iron has been used for more than 5000 years. The tips of spears and other weapons were heat treated and hardened by metallurgical processes before the word "metallurgy" was known. Apparently, some important metallurgical methods were stumbled onto accidentally and used long before people knew why they worked.

Today, technology marches forward. Our mass of knowledge has doubled in less than 50 years. It is said that more than 90 percent of the scientists who have ever lived are still alive today. It is a certainty, then, that successful students of metallurgy will find new horizons ahead in their careers.

TEST YOUR KNOWLEDGE

Write your answers on a separate sheet of paper. Do not write in this book.

1. What are some things that can be learned from a study of metallurgy?
2. What metal is the main ingredient in steel?
3. Why is steel the most popular material in the study of metallurgy?
4. What metallurgical process was used to solve the problem that involved shock forces and wear in Practical Example 1?
5. What type of problem did the metallurgical process known as "process annealing" solve in Practical Example 2?
6. When the metallurgical process known as "water quenching" caused too much distortion, what other metallurgical process was used to solve this problem in Practical Example 4?
7. Is "tempering" generally a helpful metallurgical process or does it usually present serious problems?
8. Who in industry can benefit from a study of metallurgy? Name at least five classifications of "workers."

2 METALLURGICAL AND CHEMICAL TERMS

After studying this chapter, you will be able to:

☐ State the meaning of many of the basic terms that are used in metallurgy.
☐ Tell how chemistry is related to metallurgy.
☐ Describe what a metal and an alloy are.
☐ Define chemical terms such as element, compound, solution, and atom.

There are many chemical and metallurgical terms that you need to learn in the study of metallurgy. Therefore, before you get your mind and fingers into the physical operations of heat treating, you should understand what is going on inside the metal; in fact, inside the metal crystals.

Studying the activity inside the metal can be extremely interesting. For example, iron and carbon are inside of every piece of steel, Fig. 2-1. The carbon is dissolved inside the iron. Understanding the nature of this dissolving action between iron and carbon is basic to the understanding of the rest of the story of metallurgy.

Fig. 2-1. All steels contain both iron and carbon.

TERMS TO KNOW

The following chemical and metallurgical terms will help you later in your study of the internal actions of the metals.

Element	Solution	Molecule
Metal	Solid solution	Crystal
Compound	Alloy	Grain
Mixture	Atom	

ELEMENT

An ELEMENT is a pure substance made up of just one kind of material. It is as simple as a material can be. No matter what you do to an element, you cannot change it to any other type of material. Whether you heat it, freeze it, machine it, break it, compress it, or use any other normal mechanical procedure, that element will remain the same basic material that it was when you started.

There are over 100 known and universally established elements. See Fig. 2-13. If everything on Earth were broken down into its simplest form, at least 100 different materials would be left that could not be simplified any further. Some of the more common elements include oxygen, nitrogen, chlorine, hydrogen, gold, lead, copper, iron, silver, manganese, aluminum, magnesium, and sulfur.

At room temperature, most elements are solid. Good examples are gold, iron, and lead. Several other elements are gases, such as oxygen and nitrogen. A few are normally liquids, such as bromine and mercury.

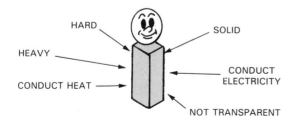

METAL

A METAL is an element that has several metallic properties. One metallic property is the ability to conduct electricity. Most metals do conduct electricity. Most metals are good conductors of heat. Metals usually are very hard. Metals are relatively heavy. Metals are not transparent. All metals have some of these properties. Some metals have every one of these metallic properties.

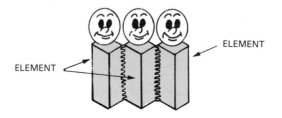

COMPOUND

A COMPOUND is a material that is composed of two or more elements that are chemically joined. A compound is not just one element. In its simplest form, it is still made up of at least two elements, Fig. 2-2.

The elements in a compound are chemically joined and, therefore, very difficult to separate. The elements stay permanently joined unless special chemical action is taken to break down the compound.

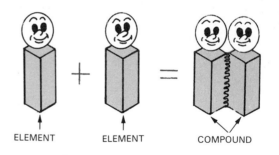

Fig. 2-2. A compound is made up of at least two elements.

One interesting feature of a compound is that its characteristics may be entirely different from the elements that make it up, Fig. 2-3. Iron sulfide is made up of iron and sulfur. Iron is magnetic, but iron sulfide is not.

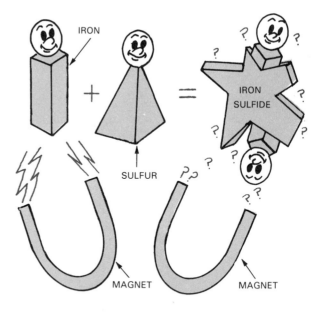

Fig. 2-3. The characteristics of a compound may be entirely different from the elements that make up the compound.

Water is made up of hydrogen and oxygen, both of which are gases. Hydrogen and oxygen will both support fires, and they are somewhat hazardous to handle. Yet, when joined together, they become water, a compound that will put out fires. See Fig. 2-4.

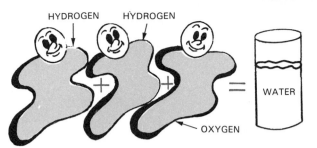

Fig. 2-4. The characteristics of water (compound) are entirely different from the characteristics of hydrogen and oxygen (elements).

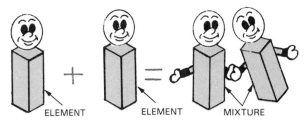

Fig. 2-6. A mixture is made up of at least two elements.

Sodium and chlorine can be chemically combined to produce table salt. Both sodium and chlorine are poisonous. Sodium is an innocent-looking, silvery metal that will burn your hand if you touch it. Chlorine is a greenish, poisonous gas that can kill you. Yet, when these two poisons are chemically combined, they become table salt, safe material to eat. See Fig. 2-5.

The difference between a mixture and a compound is the ease with which the elements can be separated. The elements in a mixture are not chemically joined; the elements in a compound are. Thus, filtering usually can separate the items that make up a mixture, Fig. 2-7.

Iron-rich vitamin tablets contain a mixture of iron and other vitamins. The iron can be removed by grinding up the tablet, then using a magnet to collect the iron particles.

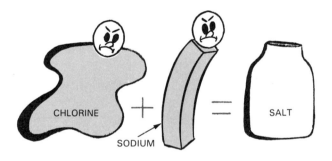

Fig. 2-5. Table salt (compound) is entirely different than sodium and chlorine (elements).

Muddy water is a mixture. In this case, a filter is not even necessary to separate the dirt from water. Just leaving the jar of muddy water stand for a period of time will permit the items in the mixture to separate.

The oil that goes into your automobile engine is a mixture of petroleum and additives. These can be separated. Homogenized milk is a mixture of milk and cream. These can be separated.

In a mixture, no component completely loses its own identity. Therefore, the characteristics of a mixture are similar to the characteristics of the items that make it up. This is another way in which a compound and a mixture are different.

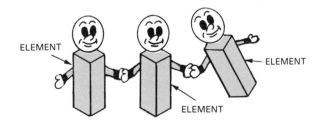

MIXTURE

A MIXTURE is a material composed of two or more elements or compounds mixed together, but not chemically joined. A mixture is not just one material. In its simplest form, it is still made up of at least two elements. See Fig. 2-6.

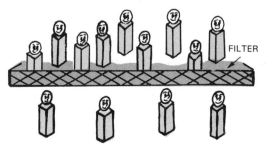

Fig. 2-7. Filtering a mixture usually separates the items that make up the mixture.

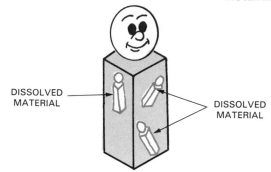

DISSOLVED MATERIAL

DISSOLVED MATERIAL

SOLUTION

A solution is a special kind of mixture. You might say that a SOLUTION is a mixture in which one substance is thoroughly dissolved in the other. When two materials combine and become a solution, one of the two will become the "dictator" and the other one will become quiet and submissive. The dictator will totally overpower the quiet one and will dissolve it. To look at a solution, you see only the dictator material, and not the dissolved material, Fig. 2-8.

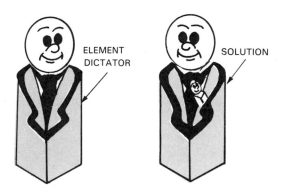

ELEMENT DICTATOR

SOLUTION

Fig. 2-8. In a solution, the dictator material is always seen.

The dictator material in a solution generally is a liquid. The dissolved material generally is either a liquid or a solid. Examples are salt water or sugar water. After sugar is dissolved in water, it is difficult to recognize the difference between sugar water and regular water. Water is the dictator. It has totally overpowered the sugar.

The dictator material is known as the SOLVENT. The dissolved material is known as the SOLUTE, Fig. 2-9. Generally, a lot more solvent is necessary than solute in order to perform the dissolving action.

The properties of a solution generally are very similar to the solvent. There will be some difference because of the influence of the solute, but not a great deal.

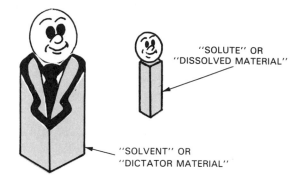

"SOLUTE" OR "DISSOLVED MATERIAL"

"SOLVENT" OR "DICTATOR MATERIAL"

Fig. 2-9. In a solution, the dictator material is known as the solvent. The dissolved material is known as the solute.

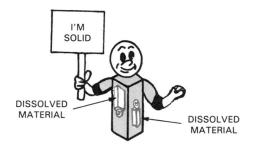

I'M SOLID

DISSOLVED MATERIAL

DISSOLVED MATERIAL

SOLID SOLUTION

A SOLID SOLUTION is a solution in which both the solvent and solute are solids. Both the dictator and the dissolved material are solids. At first, this sounds impossible. How can you mix a solid material into another solid material and cause dissolving to take place? Sugar cannot be dissolved in ice. If pieces of solid carbon are mixed up with pieces of solid iron, they will not dissolve.

The dissolving action can take place at elevated temperatures when both solids melt and become liquids, or are near their melting point. At these higher temperatures, iron will dissolve many other elements, especially carbon. Iron becomes the dictator. Small amounts of carbon or phosphorus or manganese become dissolved elements.

Copper at elevated temperatures will dissolve small amounts of zinc, lead, tin, or nickel. Many other materials behave this way, but iron and copper are two of the most common dictators.

Generally, a lot more of the dictator material — the solvent — is present than the material that is being dissolved. For example, in steel the solvent is iron. About 99 percent of the steel is iron. Only about one percent of most steels is made up of dissolved materials.

ALLOY

When two or more metals are dissolved together in a solid solution, the new material is known as an ALLOY. Steel is an alloy of iron and carbon. Bronze is an alloy of copper and tin. Brass is an alloy of copper and zinc.

The metals that are dissolved — the solutes — are also called "alloys." Thus, the word ALLOY has two meanings. 1. The dissolved metal material. 2. The solid solution that is made up of alloys and solvent.

ATOM

Back in B.C., scientists thought that if you took a piece of metal and cut it in half, both pieces would still be the same metal. This was true. They thought that if you would cut it in half again, you would still have the same metal. This was true. They also thought that if you kept cutting it into smaller and smaller pieces, you could keep this up forever and still have the same metal no matter how small those pieces became.

This was false. A metal can be cut into smaller and smaller pieces. These pieces may be so small that a microscope is required to see them. Eventually, this process reaches a limit. This limit is the atom.

The word atom means "cannot be cut." An ATOM is the smallest possible part of an element that is still that element. If an atom could be cut in half, the new pieces would no longer be that original metal.

Chemists have a theory of what the atom looks like. See Fig. 2-10. Some of this theory has been verified by x-ray diffraction techniques where tiny particles are magnified many times size. Also, the newest "electron microscope" permits scientists to faintly see the atom.

The atom is considered to be made up of a solid nucleus which contains neutrons and protons. Around this nucleus, electrons travel in

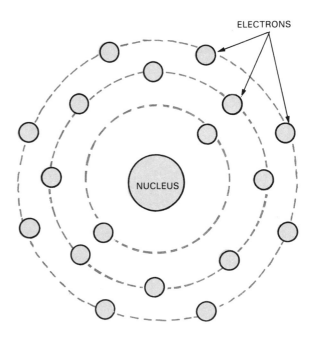

Fig. 2-10. A magnified view of the atom shows electrons and a nucleus.

circles, Fig. 2-10. There are many rings of these electrons. Each ring contains a specific number of electrons. The first ring contains two or less electrons. The second ring contains eight or less electrons. Ths third ring contains eighteen or less electrons, etc.

If an atom has few electrons, it will have few protons and neutrons in the nucleus and will be very light in weight. For example, the hydrogen atom contains only one electron. Hydrogen has one electron in its first ring, and one proton and one neutron in its nucleus, Fig. 2-11.

Helium has two electrons in its first ring, Fig. 2-12. Lithium has three electrons. Two electrons fill the first ring and the third one is left to travel in the second ring.

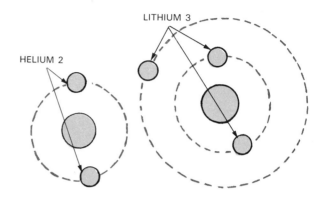

Fig. 2-12. Helium has two electrons. Lithium has three electrons.

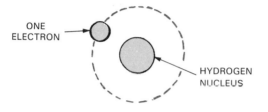

Fig. 2-11. Hydrogen has only one electron.

Each electron quantity value creates a different element, and each element has a name. The names of all elements are shown in Fig. 2-13.

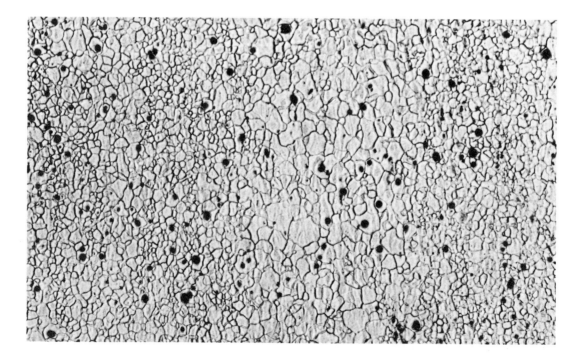

As molten metal is cooled, families of atoms or molecules form crystal structures. A microscopic picture of one of these structures is shown here.

Periodic Table of The Elements

In the periodic table the elements are arranged in order of increasing atomic number. Vertical columns headed by Roman numerals are called *Groups*. A horizontal sequence of elements is called a *Period*. The most active elements are at the top right and bottom left of the table. The staggered line (Groups IIIA-VIIA) roughly separates metallic from non-metallic elements.

Groups—Elements within a group have similar properties and contain the same number of electrons in their outside energy shell.

—The first group (IA) includes hydrogen and the alkali metals.

—The last (VIIIA) contains the *inert gases*.

—Group VIIA includes the *halogens*.

—The elements intervening between groups IIA and IIIA are called *transition elements*.

—Short vertical columns without Roman numeral headings are called sub-groups.

Periods—in a given period the properties of the elements gradually pass from a strong metallic to a strong non-metallic nature, with the last number of a period being an inert gas.

LIGHT METALS (IA, IIA) — NON METALS (IIIA, IVA, VA, VIA, VIIA)

IA	IIA											IIIA	IVA	VA	VIA	VIIA	VIIIA
Hydrogen 1.0080 H 1																	Helium 4.003 He 2
Lithium 6.939 Li 3	Beryllium 9.012 Be 4											Boron 10.811 B 5	Carbon 12.01115 C 6	Nitrogen 14.007 N 7	Oxygen 15.999 O 8	Fluorine 18.998 F 9	Neon 20.183 Ne 10
Sodium 22.990 Na 11	Magnesium 24.312 Mg 12											Aluminum 26.981 Al 13	Silicon 28.086 Si 14	Phosphorus 30.974 P 15	Sulfur 32.064 S 16	Chlorine 35.453 Cl 17	Argon 39.948 Ar 18
Potassium 39.102 K 19	Calcium 40.08 Ca 20	Scandium 44.956 Sc 21	Titanium 47.90 Ti 22	Vanadium 50.942 V 23	Chromium 51.996 Cr 24	Manganese 54.938 Mn 25	Iron 55.847 Fe 26	Cobalt 58.933 Co 27	Nickel 58.71 Ni 28	Copper 63.54 Cu 29	Zinc 65.37 Zn 30	Gallium 69.72 Ga 31	Germanium 72.59 Ge 32	Arsenic 74.922 As 33	Selenium 78.96 Se 34	Bromine 79.909 Br 35	Krypton 83.80 Kr 36
Rubidium 85.47 Rb 37	Strontium 87.62 Sr 38	Yttrium 88.905 Y 39	Zirconium 91.22 Zr 40	Niobium 92.906 Nb 41	Molybdenum 95.94 Mo 42	Technetium (99) Tc 43	Ruthenium 101.07 Ru 44	Rhodium 102.91 Rh 45	Palladium 106.4 Pd 46	Silver 107.87 Ag 47	Cadmium 112.40 Cd 48	Indium 114.82 In 49	Tin 118.69 Sn 50	Antimony 121.75 Sb 51	Tellurium 127.60 Te 52	Iodine 126.90 I 53	Xenon 131.30 Xe 54
Cesium 132.90 Cs 55	Barium 137.34 Ba 56	57-71	Hafnium 178.49 Hf 72	Tantalum 180.95 Ta 73	Tungsten 183.85 W 74	Rhenium 186.21 Re 75	Osmium 190.2 Os 76	Iridium 192.2 Ir 77	Platinum 195.09 Pt 78	Gold 196.97 Au 79	Mercury 200.59 Hg 80	Thallium 204.37 Tl 81	Lead 207.19 Pb 82	Bismuth 208.98 Bi 83	Polonium (210) Po 84	Astatine (210) At 85	Radon (222) Rn 86
Francium 223 Fr 87	Radium (226) Ra 88	89-103	104	105	106	107	108	109	110	111	112	113	114	115	116	117	118

Lanthanum 138.91 La 57	Cerium 140.12 Ce 58	Praseodymium 140.91 Pr 59	Neodymium 144.24 Nd 60	Promethium (147) Pm 61	Samarium 150.35 Sm 62	Europium 151.96 Eu 63	Gadolinium 157.25 Gd 64	Terbium 158.92 Tb 65	Dysprosium 162.50 Dy 66	Holmium 164.93 Ho 67	Erbium 167.26 Er 68	Thulium 168.93 Tm 69	Ytterbium 173.04 Yb 70	Lutetium 174.97 Lu 71
Actinium 227 Ac 89	Thorium 232.04 Th 90	Protactinium (231) Pa 91	Uranium 238.03 U 92	Neptunium (237) Np 93	Plutonium (242) Pu 94	Americium (243) Am 95	Curium (247) Cm 96	Berkelium (249) Bk 97	Californium (251) Cf 98	Einsteinium (254) Es 99	Fermium (253) Fm 100	Mendelevium (256) Md 101	Nobelium (254) No 102	Lawrencium (257) Lr 103

Fig. 2-13. A Periodic Table of The Elements lists all universally established elements by name, symbol, atomic number, and atomic weight. (Ted Bates Navy Account Group, Ted Bates & Co., Inc.)

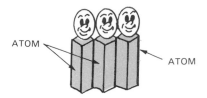

MOLECULE

When two or more elements combine to form a compound, their atoms get together and form MOLECULES. For example, a large quantity of oxygen (element) and a large quantity of hydrogen (element) combine to form a large quantity of water (compound). Thus, at the same time but on a smaller scale, two atoms of hydrogen and one atom of oxygen get together to form one molecule of water. See Fig. 2-14.

MOLECULES are not formed in a mixture, nor are they formed in a solution. Atoms of different metals may "mix" together, but they do not make a permanent chemical change in a mixture or a solution. A permanent chemical change occurs only in a compound.

The atoms in a molecule are joined together by chemical action. The atoms borrow or lend or share the electrons in their outer ring. In a molecule of water, Fig. 2-15, oxygen borrows the atoms of hydrogen to form its outer ring. If oxygen and hydrogen are just "mixed" together, there would not be sharing or giving away electrons.

The iron and carbon in steel do not take chemical action with each other. Compounds and molecules are NOT formed in steel. The atoms of iron and carbon are merely "mixed" together and become an alloy or solid solution.

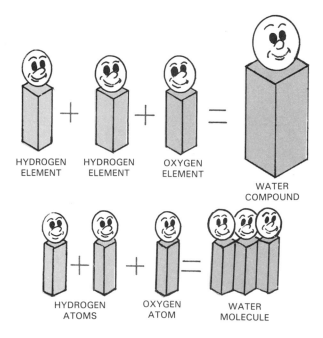

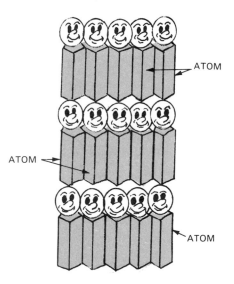

GRAINS AND CRYSTALS

When a large group of atoms or molecules get together, they form a FAMILY. These families of atoms may be large enough to be seen by the naked eye. Such a family is known as a "grain" or "crystal." In a grain or crystal, all of the atoms orient themselves in neat orderly formations.

Grains and crystals will be covered in detail in Chapter 7.

Fig. 2-14. A molecule of water is made up of two atoms of hydrogen and one atom of oxygen.

Simply put: An ATOM is the smallest part of an element. A MOLECULE is the smallest part of a compound. It takes two or more elements to make a compound. It takes two or more atoms to make a molecule.

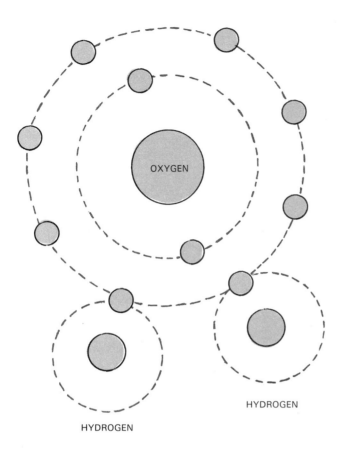

Fig. 2-15. A molecule of water is formed by chemical action between the hydrogen and oxygen atoms that make up the molecule.

APPLYING CHEMICAL TERMS TO STEEL

The chemical and metallurgical terms from this chapter that appear most often in the study of metallurgy are "crystal" and "atom" and "alloy." Steel is an alloy or solid solution. Iron is the dictator solvent. Carbon is always one of the dissolved materials.

Many other atoms or alloys are dissolved in iron to make up different special types of steel. Some of these alloys include sulfur, manganese, aluminum, phosphorus, molybdenum, tungsten, and silicon. As these atoms collect in colonies and solidify, grains or crystals are born. These terms will occur frequently throughout the study of metallurgy.

TEST YOUR KNOWLEDGE

Write your answers on a separate sheet of paper. Do not write in this book.

Use the following words to answer the 11 questions below. Some of these words may be used more than once, some of the words may not be used at all.

alloy	element	neutron
atom	grain	nucleus
compound	metal	proton
crystal	mixture	solid solution
electron	molecule	solution

1. A substance composed of two elements that are chemically joined is called _____.
2. The smallest part of a piece of silver that is still silver is called _____.
3. Name the term in which dissolving is involved, as well as a room temperature liquid.

   _____
4. Name the item that is essentially the same thing as an alloy.
5. Another name for a grain is a _____.
6. What moves around and around the nucleus in rings?
7. If an element is broken down smaller and smaller until it can no longer be broken down and still maintain the characteristics of the original piece, that smallest portion is known as _____.
8. If a compound is broken down smaller and smaller until it can no longer be broken down and still maintain the characteristics of the original piece, that smallest portion is known as _____.
9. Gold, lead, oxygen, hydrogen, nickel, helium and chlorine all are materials that cannot be broken down into anything simpler. What chemical term would a container full of one of them be known as?
10. Two elements combined but not chemically joined form a _____.
11. An element that has metallic properties is known as a _____.

3 WHAT IS STEEL?

After studying this chapter, you will be able to:

☐ Explain what steel is.
☐ Describe what the difference is in iron and steel.
☐ Point out what effect different alloys have on steel.
☐ Describe all the different kinds of steel.
☐ Compare the different kinds of cast iron.

Steel is one of the most widely used materials in the world. It has high strength. It can be machined and formed easily. Steel also is readily available and reasonably priced, compared to other materials having similar physical properties.

COMPOSITION OF STEEL

STEEL is a material composed primarily of iron. Most steel contains more than 90 percent iron. Many carbon steels contain more than 99 percent iron.

All steel contains a second element, which is carbon. Many other elements, or alloys, are contained in most steels, but iron and carbon are the only elements that are in all steel, Fig. 3-1. The

percent carbon in steel ranges from just above 0 percent to approximately 2.0 percent. Most steels have between 0.15 percent and 1.0 percent carbon.

Steels with the least carbon are more flexible and ductile (tend to deform appreciably before fracture), but they are not as strong. However, as the carbon content increases, so do strength, hardness, and brittleness.

In making steel, the iron dissolves the carbon. When there is too much carbon for the iron to "digest," the alloy is no longer called steel. The carbon precipitates out and remains as carbon flakes or other shapes. See microscopic structures in Fig. 3-2. The point where not all of the carbon can be dissolved comfortably occurs at about 2.0 percent carbon.

STEEL NUMBERING SYSTEM

Each type of steel has a name, just as human beings have names. However, a steel's name tells you a lot more about that type of steel than you would learn about Matthew or Jacob or Dawn just by hearing that person's name.

A steel's name usually consists of four numbers or digits. See Fig. 3-3. The first two digits refer to the alloy content. The last two digits (or three digits in the case of a five-digit name) refer to the percent carbon in the steel.

In 5147 steel, for example, the "51" tells you that the steel has a lot of chromium in it. In 2517 steel, the "25" indicates that there is an unusual amount of nickel in this steel. In 4718 steel, the "47" tells you that there is a more than the average

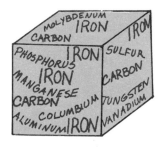

Fig. 3-1.   Iron and carbon are in all steel. Most steels also contain many other elements or alloys.

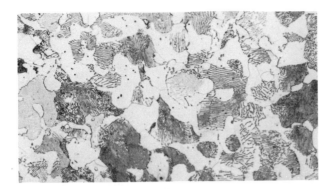

A

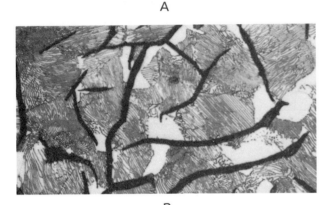

B

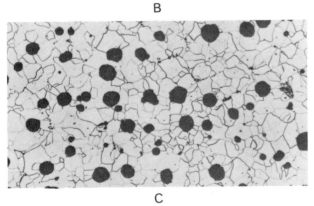

C

Fig. 3-2. Three views of microscopic structures: A — In steel, the iron dissolves the carbon. (Buehler Ltd.) B — In gray cast iron, the carbon precipitates out as carbon flakes. (Buehler Ltd.) C — In ductile cast iron, the carbon precipitates out as small round nodules. (Iron Castings Society)

ALLOY CONTENT   CARBON PERCENT

Fig. 3-3. A steel's name usually consists of four numbers and gives you information regarding the alloy content and the carbon percentage.

amount of chromium, nickel, and molybdenum in the steel.

In 1040 steel, the "10" tells you that the steel has very little alloy content except carbon. In this way, the first two numbers always give an indication of the alloy content in the steel. This steel naming system is summarized in Fig. 3-4.

Steel Numerical Name	Key Alloys
10XX	Carbon only
11XX	Carbon only (free cutting)
13XX	Manganese
23XX	Nickel
25XX	Nickel
31XX	Nickel-Chromium
33XX	Nickel-Chromium
303XX	Nickel-Chromium
40XX	Molybdenum
41XX	Chromium-Molybdenum
43XX	Nickel-Chromium-Molybdenum
44XX	Manganese-Molybdenum
46XX	Nickel-Molybdenum
47XX	Nickel-Chromium-Molybdenum
48XX	Nickel-Molybdenum
50XX	Chromium
51XX	Chromium
501XX	Chromium
511XX	Chromium
521XX	Chromium
514XX	Chromium
515XX	Chromium
61XX	Chromium-Vanadium
81XX	Nickel-Chromium-Molybdenum
86XX	Nickel-Chromium-Molybdenum
87XX	Nickel-Chromium-Molybdenum
88XX	Nickel-Chromium-Molybdenum
92XX	Silicon-Manganese
93XX	Nickel-Chromium-Molybdenum
94XX	Nickel-Chromium-Molybdenum-Manganese
98XX	Nickel-Chromium-Molybdenum
XXBXX	Boron
XXLXX	Lead

Fig. 3-4. Table relates the alloy content in steel to the first two digits of its name.

The last two digits (or three digits) indicate the percent carbon that the steel contains. In 1040 steel, for example, the "40" tells you that there is 0.40 percent carbon in the steel. In 1018 steel, the "18" indicates that there is only 0.18 percent carbon in it, thus, a very low carbon steel. An 8660 steel contains approximately 0.60 percent carbon, which would make it a medium carbon steel. See Fig. 3-5.

When the carbon content of the steel is 1.00 percent or more, three digits are needed to describe the carbon content. For example, 50100 steel contains 1.00 percent carbon.

Therefore, you can see that the four or five numbers in the name of the steel tell much about its alloy content. Indirectly, they tell you about the quality of steel, the strength of the steel, the corrosion resistance of the steel, etc. For example, 8622 steel would have .22 percent carbon, 0.20 percent molybdenum, 0.50 percent chromium, and 0.55 percent nickel, Fig. 3-6. Because of the molybdenum content, 8622 is stronger than the average steel at high temperature. Because of the chromium content, 8622 is more resistant to corrosion than the average steel. Because of the nickel content, 8622 has better toughness than the average steel. How can you tell how much carbon is in 8622 steel?

In Fig. 3-7, several more steel names are listed. The carbon content can be estimated from the last two digits. The alloy content can be estimated by using Fig. 3-4.

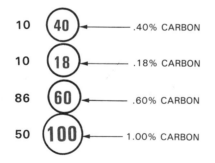

Fig. 3-5. In steel, the percent carbon can be determined from the last two or three digits.

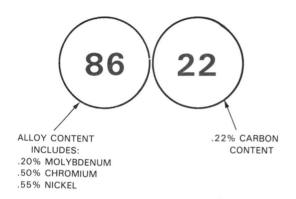

Fig. 3-6. The first two digits of a steel name tell the alloy content. The last two digits describe the carbon content.

STEEL NAME	APPROXIMATE % CARBON	ALLOYS PRESENT IN LARGER AMOUNTS THAN NORMAL CASES
1020	.20%	Only Carbon
1040	.40%	Only Carbon
1095	.95%	Only Carbon
1118	.18%	Only Carbon
1340	.40%	Manganese
2340	.40%	Nickel
2512	.12%	Nickel
3140	.40%	Nickel & Chromium
3310	.10%	Nickel & Chromium
4024	.24%	Molybdenum
4140	.40%	Chromium & Molybdenum
4320	.20%	Nickel, Chromium & Molybdenum
4620	.20%	Nickel & Molybdenum
5135	.35%	Chromium
52100	1.00%	Chromium
6150	.50%	Chromium & Vanadium
8622	.22%	Nickel, Chromium & Molybdenum
9255	.55%	Silicon & Manganese

Fig. 3-7. List gives examples that show how carbon percentage and alloy content can be recognized merely by observing the numerical name of the steel.

DIFFERENCE IN STEEL AND IRON

In Fig. 3-8, the relationship of steel to cast iron and wrought iron is shown. Note that the difference is primarily based on their carbon content. Steel ranges from just above 0 percent to approximately 2.0 percent carbon. Most cast iron contains from 2.0 to about 4.0 percent carbon. Wrought iron contains essentially no carbon. At approximately 6.0 percent carbon content, the material becomes so brittle that it is relatively useless.

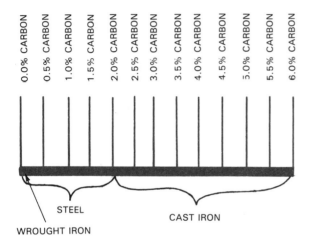

Fig. 3-8. The basic difference in wrought iron, steel, and cast iron is their percent carbon content.

EFFECT OF ALLOYS

Most steel contains other ingredients besides iron and carbon. These ingredients are commonly called ALLOYS. Most of the alloys in steel are in small amounts, but they have a great effect upon the character of the steel. Some of these alloys and their effects are summarized in Fig. 3-9.

To obtain greater strength in steel, carbon, manganese, or nickel are added. To obtain better corrosion resistance or resistance to atmospheric conditions, chromium or copper may be added. If lead or sulfur are added, the steel will have better machinability. To obtain better physical properties at high temperature, tungsten or molybdenum are recommended.

STEEL ALLOY	EFFECT ON STEEL
CARBON	Hardness — Strength — Wear
CHROMIUM	Corrosion Resistance — Hardenability
LEAD	Machinability
MANGANESE	Strength — Hardenability — More Response To Heat Treat
ALUMINUM	Deoxidation
NICKEL	Toughness — Strength
SILICON	Deoxidation — Hardenability
TUNGSTEN	High Temperature Strength — Wear
MOLYBDENUM	High Temperature Strength — Hardenability
SULFUR	Machinability
TITANIUM	Elimination Of Carbide Precipitation
VANADIUM	Fine Grain — Toughness
BORON	Hardenability
COPPER	Corrosion Resistance — Strength
COLUMBIUM	Elimination of Carbide Precipitation
PHOSPHORUS	Strength
TELLURIUM	Machinability
COBALT	Hardness — Wear

Fig. 3-9. Table lists the common alloys of steel and their effect on the steel.

The greater the amount (percentage) of the alloying elements that are added, the more profound their effect will be on the steel. However, it is unusual to have more than 2.0 percent of any one alloy. For example, phosphorous and manganese are in most steels, but the amount added rarely goes over 0.05 percent. See the table of alloys in steel in Fig. 3-10.

		EXAMPLES OF ALLOYS IN STEEL									
STEEL	TYPE OF STEEL	TENSILE STRENGTH x1000 psi	C	Mn	P	S	Si	Ni	Cr	Mo	V
1025	Plain Carbon	60-103	.22-.28	.30-.60	.04 max	.05 max					
1045	Plain Carbon	80-182	.43-.50	.60-.90	.04 max	.05 max					
1095	Plain Carbon	90-213	.90-1.03	.30-.50	.04 max	.05 max					
1112	Free Cutting Carbon	60-100	.13 max	.70-1.00	.07-.12	.16-.23					
1330	Manganese	90-162	.28-.33	1.60-1.90	.035	.040	.20-.35				
2517	Nickel	88-190	.15-.20	.45-.60	.025	.025	.20-.35	4.75-5.25			
3310	Nickel-Chromium	104-172	.08-.13	.45-.60	.025	.025	.20-.35	3.25-3.75	1.40-1.75		
4023	Molybdenum	105-170	.20-.25	.70-.90	.035	.040	.20-.35			.20-.30	
52100	Chromium	100-240	.98-1.10	.25-.45	.035	.040	.20-.35		1.30-1.60		
6150	Chromium-Vanadium	96-230	.48-.53	.70-.90	.035	.040	.20-.35		.80-1.10		.15 min
9840	Nickel-Chromium Molybdenum	120-280	.38-.43	.70-.90	.040	.040	.20-.35	.85-1.15	.70-.90	.20-.30	
4140	Chromium-Molybdonum	95-125	.38-.43	75-1.00	.035	.040	.20-.35		.80-1.10	.15-.25	

Fig. 3-10. Table shows the alloy content in several typical steels.

TYPES OF STEEL

There are many different categories and types of steel. However, most steel is classified as either carbon steel or alloy steel, Fig. 3-11. Carbon steel contains less alloys and less expensive ingredients. Therefore, carbon steel is less expensive. Alloy steels have special qualities such as increased strength, corrosion resistance, or high temperature capability.

Carbon steel can be broken down further into low carbon steel, medium carbon steel, and high carbon steel. There are many types of alloy steel such as structural steel, maraging steel, etc., Fig. 3-12. Stainless steel and tool steel are so widely used that they could be considered to be separate types of steel in themselves.

CARBON STEELS

Carbon steel comprises by far the largest tonnage of all steel. About 90 percent of all steel made is carbon steel. It is sometimes called "plain carbon steel."

CARBON STEEL includes those steels wherein carbon plays the predominant role, and there are comparatively less other alloys present in the steel. Most carbon steels are considerably less expensive than alloy steel. The three basic types of carbon steel are low carbon steel, medium carbon steel, and high carbon steel.

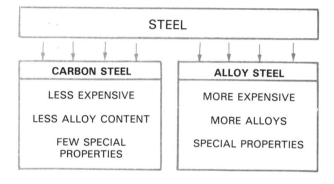

Fig. 3-11. Most steel can be broken down into one of two categories.

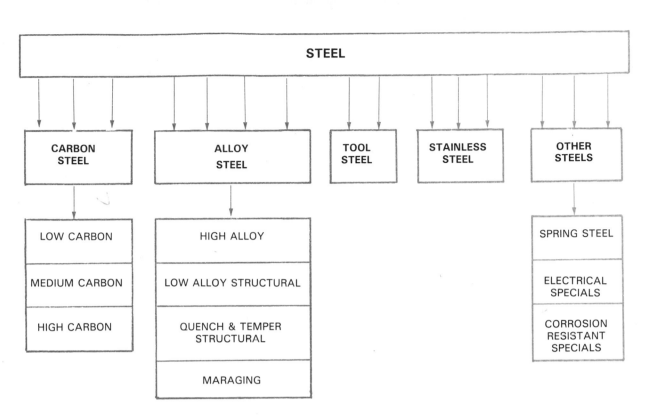

Fig. 3-12. Several types of alloy steel are so widely used that they can be considered to be categories in themselves. These include tool steel, stainless steel, and special steels.

LOW CARBON STEEL

The largest tonnage of all the carbon steel is LOW CARBON STEEL. It contains between .05 percent and 0.35 percent carbon. Low carbon steel lacks the ability to become as hard and strong as other steel. However, because it does not become hard, it is easier to machine and work with in the manufacturing plant.

Low carbon steel is the least expensive of all categories of steel. For this reason, it has many uses, Fig. 3-13. Applications include fence wire, auto bodies, galvanized sheets, storage tanks, large pipe, and various parts in buildings, bridges, and ships.

Just because low carbon steel is not as strong and hard as some of the more expensive grades of alloy steel, that should not make you think it is weak and of low quality. All steel — even

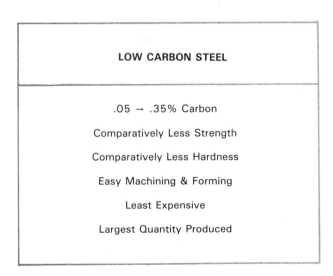

Fig. 3-13. Chart lists the special characteristics of low carbon steel.

low carbon steel — is very strong and can be trusted to support a great deal of force or weight.

MEDIUM CARBON STEEL

MEDIUM CARBON STEEL contains 0.35 percent to 0.50 percent carbon. It can be heat-treated. If heat treated properly, medium carbon steel can become quite hard and strong, Fig. 3-14. It is frequently used in forgings and high strength castings.

Applications of medium carbon steel include wheels, axles, crankshafts, and gears.

MEDIUM CARBON STEEL
.35 → .50% Carbon
Hard & Strong After Heat Treating
More Expensive Than Low Carbon Steel

Fig. 3-14. Chart gives the special characteristics of medium carbon steel.

HIGH CARBON STEEL

The carbon content in HIGH CARBON STEEL is over 0.50 percent, and it may be over 1.0 percent in some high carbon steels. These steels can be readily heat treated to obtain high strength and high hardness.

The problem that usually occurs in steels having high hardness is that they also have a relatively high distortion rate and can crack or become very brittle during hardening, Fig. 3-15. Nevertheless, high carbon steel still can be used safely in making tools, dies, knives, railroad wheels, and many other high strength applications.

HIGH CARBON STEEL
.50 → 1.00% Carbon
Hard & Strong After Heat Treating
More Expensive Than Low & Medium Carbon Steels

Fig. 3-15. Chart reveals special characteristics of high carbon steel.

ALLOY STEELS

An ALLOY STEEL is a grade of steel in which one or more alloying elements have been added in larger amounts to give it special properties that ordinarily cannot be obtained with carbon steel. See Fig. 3-16. Any steel that has an alloy content of more than the percentage listed in Fig. 3-17 would be classified as an alloy steel.

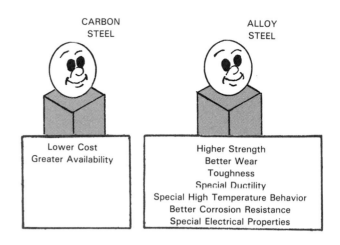

Fig. 3-16. Compare the advantages of carbon steel and alloy steel.

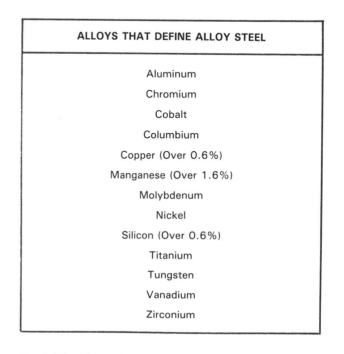

Fig. 3-17. These elements are popular alloys used in alloy steel.

Some of the special properties that the alloys may be used for are corrosion resistance, high temperature capability, electrical-magnetic behavior, high strength, and wear resistance.

Alloy steels are more expensive. They should be used only when a special property is definitely needed.

LOW ALLOY STRUCTURAL STEELS

LOW ALLOY STRUCTURAL STEELS have less alloy content than most of the rest of the alloy steels. This gives them the desirable characteristics of being stronger than carbon steels yet not much more expensive. See Fig. 3-18.

LOW ALLOY STRUCTURAL STEEL
Less Alloys Than Other Alloy Steels
More Alloys Than Carbon Steels
Less Expensive Than Other Alloy Steels
More Expensive Than Carbon Steels
Weldable
Structural Applications
Alloys Include: Manganese-Silicon-Columbium-Vanadium-Copper
Good Corrosion Resistance

Fig. 3-18.   Chart lists the special characteristics of low alloy structural steel.

Fig. 3-19.   More than 22,000 tons of structural steel was used during construction of the Pacific Gas and Electric Company Headquarters Building in San Francisco. This building is 34 stories high, contains more than 1,000,000 sq. ft. of floor space and rises 530 ft. above the street. (Bethlehem Steel Corporation)

Low alloy structural steels contain more than the average amount of manganese, silicon, columbium, vanadium, and copper. These steels are used primarily for structural applications where weight-saving is important. Because they are stronger than carbon steels, low alloy structural steel parts can be made smaller, thus saving weight. Figs. 3-19 through 3-21 show applications of different types of structural steel.

QUENCH AND TEMPER STRUCTURAL STEELS

QUENCH AND TEMPER STRUCTURAL STEELS are stronger than the low alloy structural steels and have better impact resistance at lower temperatures. These steels also have better corrosion resistance and better characteristics in general than low alloy structural steels. See Fig. 3-22.

Quench and temper structural steels are often used in pressure vessels, submarine bodies, and other applications where additional cost is justified in order to obtain greater strength and corrosion resistance.

MARAGING STEELS

MARAGING STEELS contain high amounts

Fig. 3-20. Madison Square Garden Sports and Entertainment Center, New York City, is shown under construction. Here, 48 assemblies of 3 3/4 inch steel strand cable were manufactured by Bethlehem Steel Corporation to support the structure's 404 ft. diameter roof. This cable-supported roof is one of the largest permanent suspension roofs in the United States. (Bethlehem Steel Corporation)

Fig. 3-21. Different types of structural steel were used in construction of the Golden Gate Bridge. (American Iron and Steel Institute)

QUENCH & TEMPER STRUCTURAL STEEL
Stronger Than Low Alloy Structural Steel
Better Properties Than Low Alloy Structural Steel
More Expensive Than Low Alloy Structural Steel
Structural Applications

Fig. 3-22. Chart gives the special characteristics of quench and temper structural steel.

of nickel and small amounts of carbon. Generally, 18-25 percent of nickel is found in maraging steels. They have a very high strength, up to 250,000 psi, but still maintain ductility and good toughness. Some maraging steels can be stretched 11 percent before breaking, Fig. 3-23.

MARAGING STEEL
18-25% Nickel
Low Carbon Content
Very High Strength
Good Ductility & Toughness

Fig. 3-23. Chart shows special characteristics of maraging steel.

Maraging steels are used in rocket motor cases and other aerospace applications wherein both high strength, good toughness, and ductility are necessary.

A typical alloy composition of maraging steels is shown in Fig. 3-24.

Fig. 3-25. This piercing die is made from Carpenter Vega® a commercial brand of A6, air hardening tool steel. (Carpenter Technology Corporation, Reading, PA)

TYPICAL COMPOSITION OF MARAGING STEELS	
CARBON	.03%
NICKEL	18.5%
COBALT	7.5%
MOLYBDENUM	4.8%
TITANIUM	0.4%
ZIRCONIUM	.01%
ALUMINUM	.10%
SILICON	.10% max
MANGANESE	.10% max
SULFUR	.01% max
PHOSPHORUS	.01% max

Fig. 3-24. Chart lists the typical composition of maraging steels.

TOOL STEELS

TOOL STEELS are so widely used that they are considered to be a separate steel category, Fig. 3-12. There are many different types of tool steel. Certain types are used in cutting tools, molds, and dies, Figs. 3-25 through 3-27. Tool steels are also used for general machine parts, where high strength, wear resistance, and dimensional stability are required.

Fig. 3-26. This tool and die maker is polishing a slide insert of a Ford transmission case die. It is made of Crucible Nu-Die® V Densified™, a commercial brand of H-13 tool steel. (Crucible Specialty Metals Division of Colt Industries)

Fig. 3-28 shows eleven common categories of tool steel. Each of these categories denotes a tool steel that is used for a different purpose.

The "S" category of tool steel, for example, is used for extreme shock resistance applications, such as air hammers or stamping dies.

The "A" series of tool steels have special alloys in them that make them capable of becoming hard when quenched in air. Air quenching is a less violent method of quenching than water quenching or oil quenching. "A" series tool steels will harden by water quenching or oil quenching or air quenching. However, quenching them in air makes them less likely to crack or distort during the quenching process. (See Chapter 11).

Therefore, these more expensive "A" category tool steels are used where dimensional accuracy is extremely important.

The "M" and "T" types of tool steels are for high-speed work and contain higher quantities of molybdenum and tungsten.

The "H" category of tool steel has good high temperature strength. The "H" types are used for hot working processes such as forging and die casting.

STAINLESS STEEL

Stainless steel is a category of special alloy steel that is used extensively, Fig. 3-12. As the name

Fig. 3-27. This die maker demonstrates the casting position of the prototype in the stationary half of a Ford transmission case die. The die weighs more than 50 tons. It contains about 20 inserts made of H-13 tool steel.
(Crucible Specialty Metals Division of Colt Industries)

TOOL STEEL CATEGORIES		
CATEGORY	EXAMPLE	DESCRIPTION
W	W1 & W5	Water Hardening
O	O1 & O6	Oil Hardening
A	A2 & A6	Air Hardening
D	D1 & D2	Oil or Air Hardening
S	S2 & S4	Shock Resisting
H	H10 & H41	Hot Working
M	M1 & M34	High Speed (Molybdenum)
T	T2 & T15	High Speed (Tungsten)
L	L1 & L2	Special Purpose
F	F1 & F3	Special Purpose
P	P5 & P20	Mold Making

Fig. 3-28. There are 11 common categories of tool steel.

implies, stainless steel has extremely good corrosion resistance. It costs more than carbon steel and is harder to cut and machine. However, if a machined part is expected to come into contact with a corrosive atmosphere, the extra cost is well justified. When there are special sanitation requirements such as in food processing, or in the transfer of chemicals through pipes, stainless steel is commonly used.

All stainless steels contain high quantities of chromium alloy, and many contain high quantities of nickel alloy. These alloys give stainless steel its superior ability to resist corrosion. See Fig. 3-29 through 3-31.

Fig. 3-30. Stainless steel is used for building trim on this structure. (Armco Inc.)

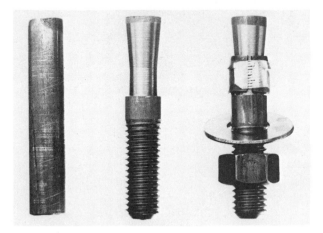

Fig. 3-29. This stainless steel KWIK-BOLT® Concrete Anchor is made in three production stages. 1 — 1/2 in. round bar stock of 303 PLUS-X stainless steel. 2 — Part has been machined and degreased 3 — KWIK-BOLT® complete with floating wedges, thermoplastic sleeve, and foil retainer.
(Crucible Specialty Metals Division of Colt Industries)

Fig. 3-31. Stainless steel is used as trim around these windows. (Armco Inc.)

SPRING STEEL

SPRING STEEL is a special category of steel that has great hardness, strength, and elasticity. Its uses include leaf springs, clock springs, knife blades, and golf club shafts. See Fig. 3-32.

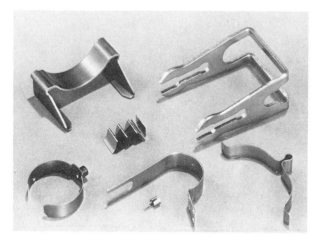

Fig. 3-32. Spring steel is used to make a variety of spring-like parts. (Wallace Barnes Steel, Barnes Group Inc.)

Spring steel is normally purchased as a thin, flat material. The carbon content can vary anywhere from 0.35 to 1.4 percent.

Compared with carbon steel, spring steel is relatively expensive. However, it has special strength and ductility characteristics that make it worth the extra cost. Additional alloys in spring steel include manganese up to .8 percent, and chromium, silicon, vanadium, or molybdenum.

Applications of spring steel include: camera shutters, circular cutters in the machine tool industry, steel used in fishing rods, putty knives, manicure files, bread knives, the reeds used in electronic vibrators, measuring tapes, steel rules, trowels for cement work, feeler inspection gauges, and innumerable small springs for clocks and other applications. See Fig. 3-33.

OTHER HIGH ALLOY STEELS

There are many other special purpose steels that have unusual alloy contents.

1. For the electrical industry, special steels with high silicon content are used in generators and transformers.
2. In rockets, missiles, jet aircraft, and nuclear power devices, where high temperature is important, special alloy steels that contain titanium, columbium, nickel, and chromium have been developed.
3. Special cobalt steels that contain over 30 percent cobalt are used for electromagnets.

Many of these special purpose steels are extremely expensive and can be justified only when their unusual properties are necessary for safety or special needs.

Fig. 3-33. Spring steel is made into many different shapes for a variety of applications other than just conventional springs. (Wallace Barnes Steel, Barnes Group Inc.)

CAST IRON

CAST IRON IS A MATERIAL containing primarily IRON, 2.0 to 6.0 percent CARBON, and small amounts of SILICON. Other ALLOYS are usually there. Bear in mind that "cast iron" and "iron" are two totally different terms. Consider it a coincidence that the word "iron" is in both titles.

HOW STEEL AND CAST IRON DIFFER

First, however, we need to clarify the difference between steel and cast iron. In making steel, the carbon that is added to the iron dissolves and disappears. This is similar to the dissolving action of sugar in water. If you added a little sugar to a glass of water, the sugar would immediately dissolve and be invisible. If you added a little more sugar, the same thing would happen. Eventually, if you added enough sugar, the water could not dissolve any more and some sugar would precipitate out and you could see it.

This same situation exists with iron and carbon in steel and cast iron. If you put carbon in (hot) iron, the carbon will dissolve and disappear. Eventually, if you put in too much carbon, it will precipitate out. Thus, steel is iron with the carbon still in solution, which occurs below 1.6 to 2.0 percent (the percentage varies in different steels). Cast iron is iron in which some of the carbon has precipitated out and appears as flakes (most common) or little spheres. See Figs. 3-34 and 3-35.

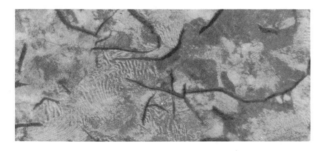

Fig. 3-34. In gray iron, carbon flakes tend to crystallize out of the iron.
(Central Foundry, Division of General Motors Corporation)

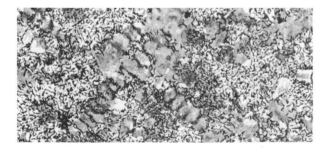

Fig. 3-35. The graphite flakes in this cast iron sample are very fine and can barely be recognized. The flakes are the tiny black lines which appear throughout the photograph. (Buehler Ltd.)

The effect of these particles that separate themselves from the iron can be both good and bad. They tend to give a cushioning action to the iron when it receives high forces that try to compress it, Fig. 3-36. This makes cast iron a very good material when vibration is present along with large compressive loads. See Fig. 3-37.

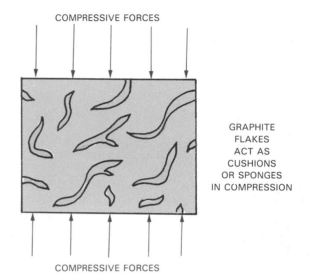

Fig. 3-36. The carbon flakes tend to give a cushioning action to the iron during compression.

On the other hand, cast iron is more brittle than steel, and it has very poor strength when stretched in tension, Fig. 3-38. The flake particles tend to encourage cracks and cause breakage to occur.

Cast iron is very easy to machine. It also is easy to cast because it has less alloy content (with the exception of carbon) than steel. Cast iron casting is also easier to control than steel casting.

APPLICATIONS

Cast iron is used extensively for the frames of large equipment and machine tools, Fig. 3-39. Its damping capacity and compressive strength make it well suited for these applications. Because cast iron has very good wear characteristics, it is used for engine blocks, piston rings, brake drums, rolls, and crushers.

Fig. 3-37. This base of a precision tool grinder absorbs vibration. Making it of cast iron gives the base a long life of service. (Iron Castings Society)

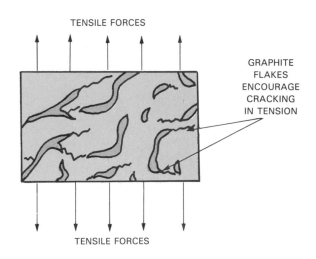

Fig. 3-38. The graphite flakes in cast iron make it weaker in tension than in compression.

In architecture, cast iron is used for stair treads and lamp posts. Cast iron is also used for manhole covers, furnace grates, and any other applications wherein its castability, machinability, damping characteristics, and compressive strength are of benefit.

TYPES OF CAST IRON

There are five basic types of cast iron: gray, white, malleable, ductile, and special alloy. See Fig. 3-40.

GRAY CAST IRON

Gray cast iron is the most widely used of all five types of cast iron. In fact, more gray cast iron is used than the other four types combined.

Although gray cast iron does not have some of the good qualities of ductile, malleable or special alloy cast iron, it is considerably less expensive. The cost differences and quality differences all must be considered when choosing which type of cast iron is to be used in a given application.

GRAY CAST IRON is very hard and brittle and has relatively poor tensile strength because of the graphite flakes in its structure. However,

Fig. 3-39. Iron castings are extensively used in paper mill equipment, such as this large pulp refiner. Cast iron meets rugged service and corrosion-resistance requirements. (Iron Castings Society)

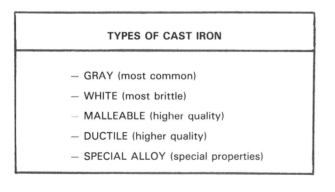

TYPES OF CAST IRON
— GRAY (most common)
— WHITE (most brittle)
— MALLEABLE (higher quality)
— DUCTILE (higher quality)
— SPECIAL ALLOY (special properties)

Fig. 3-40. Chart lists the five basic types of cast iron.

it has excellent compressive strength, damping capacity, and is easily cast, Fig. 3-41.

When people refer to a material as "cast iron," but do not specify which type, chances are they are talking about "gray cast iron."

WHITE CAST IRON

WHITE CAST IRON is not used as extensively

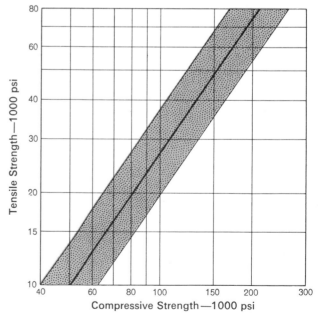

Fig. 3-41. Gray cast iron has better compressive strength than tensile strength. This graph shows the general relation between tensile strength and compressive strength in gray iron. (Iron Castings Society)

as gray cast iron because it is harder, more brittle and more difficult to machine. It also has poorer impact strength.

The extreme hardness of white cast iron, however, makes it a valuable material for some applications. For example, white cast iron rolls, used in mills, need that great hardness for breaking up stone and other materials. See Fig. 3-42.

White cast iron is also used as an intermediate step for producing malleable cast iron.

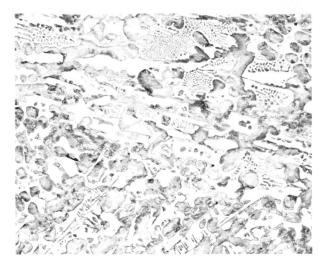

Fig. 3-43. Microscopic view of white iron. Magnification is 100X. (LECO Corporation)

Fig. 3-42. White iron is used in these muller tires because of its superior hardness and abrasion resistance. (Iron Castings Society)

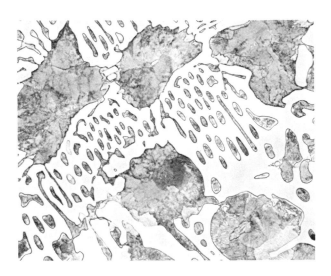

Fig. 3-44. Microscopic view of white iron. Magnification is 400X. (LECO Corporation)

When white cast iron is observed under a microscope, it looks much different than gray cast iron. See the views of the microscopic structures in Figs. 3-43 and 3-44.

MALLEABLE CAST IRON

Malleable cast iron has several special properties that make it superior to both gray cast iron and white cast iron, Fig. 3-45. Malleable cast iron has more tensile strength, ductility, and greater impact strength. It also costs more than gray cast iron or white cast iron.

To make malleable cast iron, white cast iron must first be produced. The white cast iron is then heated extensively at high temperatures to refine it. Eventually, carbides break down into carbon and free iron. The iron collects in small balls which cause the malleable cast iron to be more ductile and workable, without sacrificing high compressive strength.

Fig. 3-46 shows some typical examples of the applications of malleable cast iron.

PROPERTIES OF CAST IRON				
TYPE OF CAST IRON →	GRAY	WHITE	MALLEABLE	DUCTILE
WEIGHT ''lbs/in³''	.25-.27	.27-.28	.26-.27	.25-.27
TENSILE STRENGTH ''psi''x10⁻³	20-70	20-50	50-100	60-150
COMPRESSIVE STRENGTH ''psi''x10⁻³	100-170	100-150E	200-290	120-300
IMPACT STRENGTH V-NOTCHED CHARPY ''ft-lb''	LOW	3-10	14-17	2-30
HARDNESS ''BHN''	140-290	300-580	110-270	140 330
MODULUS OF ELASTICITY ''psi''x10⁻⁶	12-20	LOW	25-28	18-25
COEFFICIENT OF THERMAL EXPANSION ''in/in ºF'' x10⁺⁶	6	5	6-8	6-10

Fig. 3-45. Chart shows comparison of properties of the different types of cast iron.

Fig. 3-46. Typical examples of the applications of malleable cast iron. (Central Foundry, Division of General Motors Corporation)

DUCTILE CAST IRON

DUCTILE CAST IRON is sometimes referred to as nodular cast iron because its graphite is in the shape of tiny spheres. See Fig. 3-47. Ductile cast iron, as its name implies, has great ductility. Its tensile strength is higher than any other form of cast iron. Compare values of properties

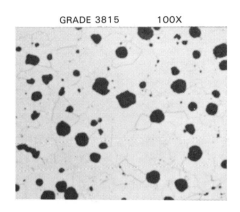

GRADE 3815   100X

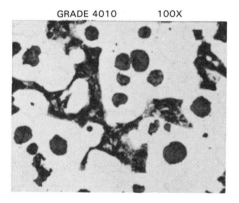

GRADE 4010   100X

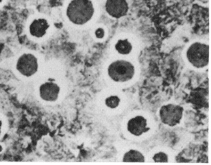

GRADE 5203   100X

Fig. 3-47. Nodular iron or ductile cast iron has graphite in the shape of tiny spheres or nodules. It has good ductility and tensile strength.
(Central Foundry, Division of General Motors Corporation)

in Fig. 3-45. As the price of ductile cast iron gradually becomes more competitive, it is replacing gray cast iron in many applications.

Figs. 3-48 and 3-49 show examples of some typical applications of ductile cast iron.

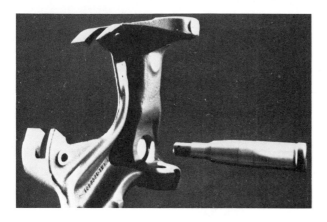

Fig. 3-48. Wheel spindle and support are ductile cast iron parts. (Central Foundry, Division of General Motors Corporation)

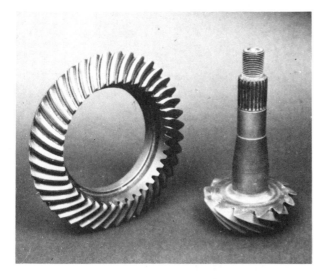

Fig. 3-49. Automotive ring gear and pinion are typical ductile cast iron parts. (Central Foundry, Division of General Motors Corporation)

SPECIAL ALLOY CAST IRON

Most cast iron is very basic in its alloy con-

tent. Cast iron primarily contains just carbon and silicon as its alloys. A number of SPECIAL ALLOY CAST IRONS have been developed for special purposes or specific applications.

A few grades of SPECIAL ALLOY cast iron contain high percentages of nickel, copper, chromium, and other alloys. The nickel, copper, and chromium give good corrosion resistance and good chemical resistance to acids. Also, some special cast iron alloys have been developed which have greater strength and better high temperature properties, Fig. 3-50.

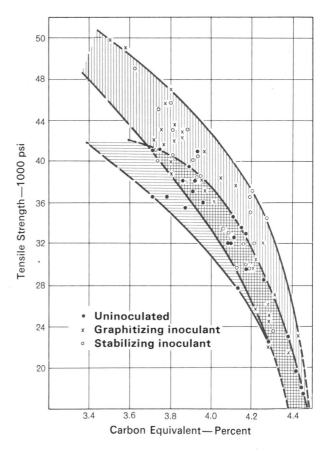
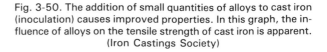

Fig. 3-50. The addition of small quantities of alloys to cast iron (inoculation) causes improved properties. In this graph, the influence of alloys on the tensile strength of cast iron is apparent. (Iron Castings Society)

Special alloy cast iron is used in cylinders, pistons, piston rings, and turbine stator vanes, Fig. 3-51.

Fig. 3-51. The vanes of this turbine stator are integrally cast of a high alloy cast iron. They are not machined. (Iron Castings Society)

WROUGHT IRON

Wrought iron is very different from cast iron. WROUGHT IRON is almost pure iron; it has very little carbon. Review Fig. 3-8.

Since wrought iron contains little carbon, it has very low strength and hardness. However, it is very ductile and corrosion resistant. Before 1860, wrought iron was the most important structural metal available. However, due to advances brought about by metallurgical research, wrought iron has been widely replaced as a structural material.

Wrought iron's key asset today is good corrosion resistance. Many fibrous stringers of slag are distributed throughout wrought iron. If corrosion attacks the iron, the deterioration proceeds only until it meets a slag stringer. The corrosion stops there, can corrode no farther, and it forms

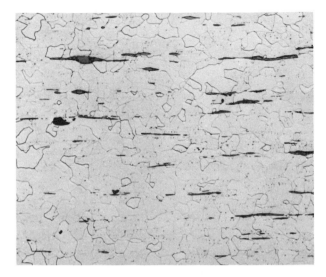

Fig. 3-52. Wrought iron contains many fibrous, elongated stringers of slag which are composed largely of FeO and SiO_2.

a protective coating. The slag then becomes a barrier against further corrosion. Fig. 3-52 shows a microscopic view of these slag stringers, which have considerable silicon in them.

A typical chemical makeup of wrought iron is shown in the chart in Fig. 3-53.

The amount of wrought iron used today is small, compared to cast iron or steel. Applications include processing tanks, waste water pipes, sheeting on ships, and gratings.

WROUGHT IRON TYPICAL CHEMICAL MAKEUP	
IRON	OVER 99.6%
CARBON	.06-.08%
SILICON	.10-.16%
MANGANESE	.02-.05%
SULFUR	.01%
PHOSPHORUS	.06-.07%

Fig. 3-53. Chart gives typical chemical makeup of wrought iron.

TEST YOUR KNOWLEDGE

Write your answers on a separate sheet of paper. Do not write in this book.

1. All steel contains _____ and _____.
2. What percent carbon is in steel?
3. What percent iron is in steel?
4. As carbon content increases in steel, does the steel become more brittle or more ductile?
5. As the carbon content increases in steel, does steel become harder or softer?
6. As the carbon content increases in steel, does steel become stronger or weaker?
7. If the name of a steel contains four digits, what do the first two digits tell you about the steel?
8. If the name of a steel contains four digits, what do the last two digits tell you about the steel?
9. Why do some steels have five digit names?
10. In 4024 steel, there is an unusual amount of _____ alloy in steel.
11. In 4147 steel, there is an unusual amount of _____ and _____ alloy in the steel.
12. In 52100 steel, there is an unusual amount of _____ alloy in the steel.
13. In 8630 steel, there is an unusual amount of _____, _____, and _____ alloy in the steel.
14. If iron contains more than 2.0 percent carbon, it is no longer called steel. It is called _____.
15. If iron contains almost no carbon, it is no longer called steel. It is called _____.
16. Name four ways in which alloys can improve steel.
17. Most steel is classified as either carbon steel or alloy steel. Which of the two is generally less expensive?
18. Which has the greatest usage today, carbon steel or alloy steel?
19. The three basic types of carbon steel are _____, _____, and _____.
20. What special properties do some alloy steels have?
21. What is a maraging steel?
22. A steel with a name like W-1 or A-6 is a type of _____ steel.
23. What does the S stand for in a steel that is named S-1?
24. The alloys that are in all stainless steels are _____, _____, and _____.
25. What practical applications can spring steel be used for besides springs?
26. Compared to steel, cast iron has poor _____ strength, but excellent _____ strength.
27. Name the five basic types of cast iron.
28. Which type of cast iron is the most commonly used?
29. What is the biggest disadvantage of white cast iron?
30. What is the advantage of gray and white cast iron over malleable and ductile cast iron?
31. What is the main advantage of malleable and ductile cast iron over gray and white cast iron?
32. Wrought iron offers very good resistance against _____ .

4 MANUFACTURE OF IRON AND STEEL

After studying this chapter, you will be able to:

☐ Explain how steel is made.
☐ Describe how cast iron is made.
☐ Point out what iron ore and pig iron are.
☐ Discuss how the rolling mill changes big ingots of steel into different shapes.

The manufacture of steel, is done in two basic steps, Fig. 4-1. Two furnaces — or metallurgical processes — are required. First, IRON ORE is converted to PIG IRON in the BLAST FURNACE, Fig. 4-2. Then, PIG IRON is made into STEEL in a STEEL-MAKING FURNACE. See Fig. 4-3.

Fig. 4-2. This huge blast furnace located at the Armco Inc. Ashland Works stands 234 ft. high with a hearth diameter of 33 1/2 ft. It has its own computer which selects the materials for the charge.

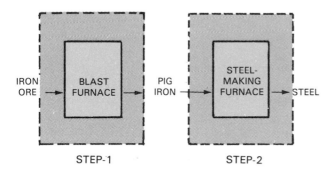

Fig. 4-1. The two basic steps of the steel-making process convert iron ore to pig iron and pig iron to steel.

THE STEEL-MAKING PROCESS

The entire process for making steel is diagrammed in Fig. 4-4. In the first step, IRON ORE is mined from the ground and shipped to one of the great steel-making centers in the United States. There, the iron ore is mixed with coke, limestone, and hot gases inside a blast furnace. The products coming out of the blast furnace are pig iron, slag, and hot gases. The PIG IRON is used to make STEEL or CAST IRON.

STEEL-MAKING FURNACES

There are three main types of steel-making furnaces:
1. Basic oxygen furnace.
2. Open hearth furnace.
3. Electric arc furnace.

In the steel-making process, the pig iron enters one of these three furnaces along with fuel, alloys, scrap steel, limestone, and small amounts of iron ore.

Fig. 4-3. In the steel-making process, an overhead crane pours a ladle of molten iron into a basic oxygen furnace. The BOF can manufacture 300 tons of steel in less than an hour.

CAST IRON-MAKING FURNACES

There are two types of cast iron-making furnaces:
1. Cupola.
2. Electric induction furnace.

In this process, the pig iron goes into one of these furnaces and is joined by fuel, limestone, and steel scrap. The fresh cast iron normally is poured directly into molds for castings.

Thousands of companies throughout the United States have cast iron furnaces. They pour final castings in their foundries. *Is there a cast iron foundry in or near your hometown?*

STEEL FOUNDRIES

While cast iron foundries are common, there are very few small steel foundries. Most of the steel-making furnaces are located near Chicago, Pittsburgh, Bethlehem (Pennsylvania), or one of the other great steel-making centers.

The steel that is poured from a steel-making furnace is just at the beginning of its long journey. Next it travels to the ingot building, then to the rolling mill. There the steel is made into its final commercial form.

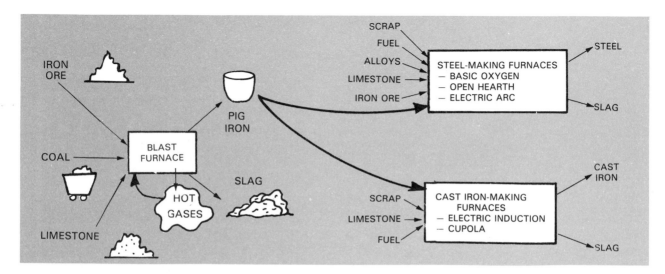

Fig. 4-4. Making steel and making cast iron involve two basic processes in each case.

Steel making and cast iron making are exciting businesses. Each of the steps shown in Fig. 4-4 will be discussed in more detail later in this chapter.

IRON ORE

Iron ore is mined from the ground. It is an unimpressive looking rock, Fig. 4-5, that may contain only 30 percent iron. The iron in the rock is in chemical compounds or ores where the iron is combined with oxygen or sulfur, Fig. 4-6. This ore is found intermixed in rock, gravel, clay, sand, or even mud, Fig. 4-7. Getting the pure iron out of this form is very costly, but necessary.

Fig. 4-5. Iron ore is a rock with high iron content. (Erie Mining Company, managed by Pickands Mather & Co., Cleveland, Ohio)

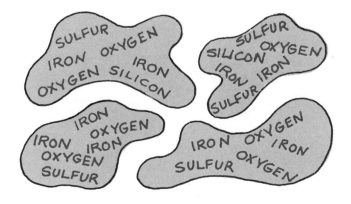

Fig. 4-6. The iron in iron ore is chemically combined with oxygen or sulfur.

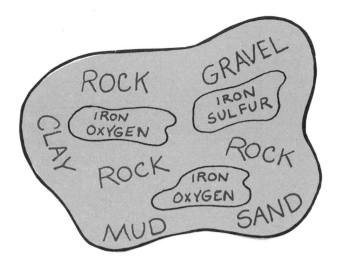

Fig. 4-7. Iron ore is found intermixed in rock, gravel, clay, sand, or mud.

IRON ORE DEPOSITS

Iron ore deposits are found in several locations in the United States. The largest concentration is in the Lake Superior region, Fig. 4-8. Most of the iron ore is mined from open pit mines. Here the rock is broken up by explosive charges, then loaded with power shovels onto trucks, Fig. 4-9, or railroad cars. It is then taken to a nearby processing plant.

IRON ORE PROCESSING

At the processing plant, Fig. 4-10, the ore goes through many operations. First, the large chunks of stone are broken up by giant CRUSHERS, Fig. 4-11. These smaller pieces are then fed into ROLL MILLS and BALL MILLS, as shown in Figs. 4-12 and 4-13. The mills rotate. Hardened steel rollers or balls rotate with the ore and steadily pound it into finer particles.

After being crushed and milled, the rock particles become so fine that they feel like sugar or dry sand. In this condition the valuable iron ore particles can be separated from the worthless particles. This is done with MAGNETIC SEPARATORS, Fig. 4-14. Strong magnets pull the iron ore particles away from the rest of the powder, as the separator rotates.

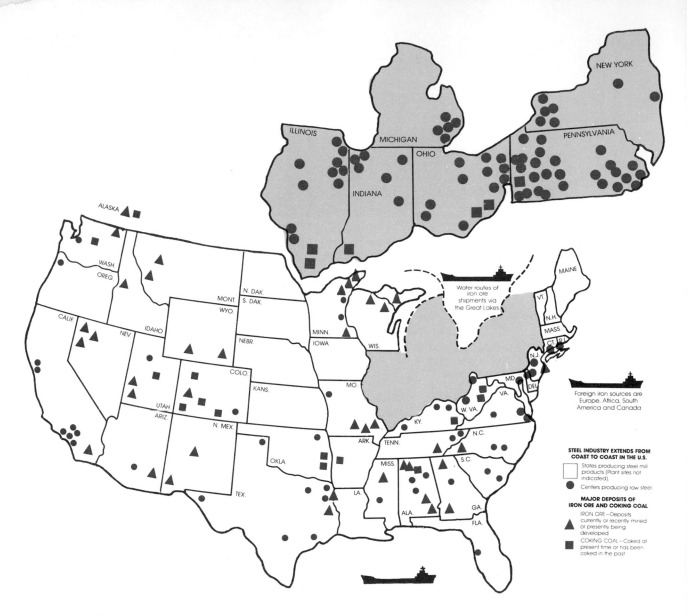

Foreign iron sources are
Europe, Africa, South
America and Canada

Water routes of
iron ore
shipments via
the Great Lakes

Fig. 4-8. Study the geography of iron and steel in the United States. (American Iron and Steel Institute)

Fig. 4-9. Power shovel loads large chunks of iron ore onto trucks at the iron ore mine.
(Erie Mining Company, managed by Pickands Mather & Co., Cleveland, Ohio)

Fig. 4-11. The iron ore is broken into smaller pieces by large crushers. (Erie Mining Company, managed by Pickands Mather & Co., Cleveland, Ohio)

Fig. 4-10. Chunks of iron ore arrive at the processing plant where the ore will go through many operations. (The Cleveland-Cliffs Iron Company)

Fig. 4-12. These ball mills break the iron ore into still smaller pieces. (Erie Mining Company, managed by Pickands Mather & Co., Cleveland, Ohio)

Fig. 4-13. Following grinding in giant primary autogenous grinding mills, the ore is further ground until it reaches a powder-fine consistency in these 15 1/2 ft. diameter by 30 ft. long pebble mills. (The Cleveland-Cliffs Iron Company)

Fig. 4-14. The iron particles are recovered by these magnetic separators. (The Cleveland-Cliffs Iron Company)

Next, the powder goes to FLOTATION CELLS where the particles are immersed in a liquid, Fig. 4-15. Impurities bubble to the top and are skimmed away, leaving an iron rich mixture. After magnetic separation and flotation, the ore may contain as much as 70 percent iron. It is then ready to be shipped to a giant steel mill.

TYPES OF IRON ORE

There are many different types of iron ore, Fig. 4-16. One of the most valuable types is called HEMATITE, which has a reddish color. This is a valuable ore because it contains a high percentage of iron.

IRON ORE	IRON COMPOUNDS IN THE ORE
Hematite	Fe_2O_3
Magnetite	Fe_3O_4
Siderite	$FeCO_3$
Pyrite	FeS
Limonite	Fe_3O_4
Taconite	Various
Jasper	Various

Fig. 4-16. This chart lists the more common types of iron ore.

Fig. 4-15. Flotation cells generate air bubbles. Impurities adhere to the bubbles and float over the top of the cell while the iron particles are drawn off the bottom. (The Cleveland-Cliffs Iron Company)

One of the most interesting types of iron ore is called TACONITE. At one time, taconite was considered worthless because it contained only 20 to 30 percent iron. Getting iron out of the ore required processing that was too expensive and time consuming. Today, due to improvements in processing methods, this taconite is one of the most popular ores. Due to modern mining technology, the refining of taconite is now more efficient.

After the taconite ore has been milled to a fine powder, and separated from the waste powder, it is poured into a BALLING DRUM. See Fig. 4-17. Here, it is mixed with a binder or "glue" and rolled into small round pellets about one-half inch in diameter. These pellets are baked and hardened. Then, they are shipped to steel centers, Fig. 4-18. It is much more convenient to handle the ore in this pellet form than when it is an iron ore powder.

BLAST FURNACE

As mentioned earlier, the blast furnace converts iron ore to pig iron. At the blast furnace, the iron ore, Fig. 4-19, meets the coke, and the limestone, Fig. 4-20. These three ingredients are

carried in skip cars, Fig. 4-21, to the top of the blast furnace. These skip cars look like little railroad cars riding on a very steep, inclined track.

Fig. 4-17. Taconite pellets are formed in a balling drum. (The Cleveland Cliffs Iron Company)

Fig. 4-18. Taconite pellets are stockpiled at Erie Mining Company processing plant. These pellets, with an iron content of about 65 percent, are comparable to high grade ore. (Bethlehem Steel Corporation)

Fig. 4-19. An enormous pile of iron ore feeds a pair of high capacity blast furnaces. (Armco Inc.)

A blast furnace, Fig. 4-22, may be more than 250 ft. high, taller than a 12-story building. The inside is about 30 ft. wide. The outside surface of the blast furnace is a thick steel shell. The inside wall is lined with fire resistant brick.

As much as 350 tons of molten iron are produced and tapped from the furnace every three to five hours. Over 80 million tons of molten iron is produced annually.

A blast furnace in operation is spectacular to see, especially at night, Fig. 4-23. The tremendous heat needed to melt the coke, limestone, and iron ore is attained by a continuous blast of hot air, introduced at the bottom of the furnace. The coke burns first. It, in turn, melts the iron ore and limestone, and a chemical reaction takes place that produces free iron.

As everything melts, it trickles down through the tall stack and collects in a molten pool at the bottom of the blast furnace.

Since the iron is the heaviest, it sinks to the bottom of the furnace. The limestone reacts with impurities, forming SLAG (a relatively non-useful byproduct). This slag floats on top of the molten pool and looks like a scum.

The hot gases work their way to the top of the blast furnace and are collected, after being filtered by dust catchers and cleaned by special equipment. These hot gases are then routed through one or more STOVES. The stoves are tall, thin, and cylindrical in shape, as shown at far left in Fig. 4-24. They may be 120 ft. high and 28 ft. in diameter. They look like junior-sized blast furnaces.

Fig. 4-20. A huge bucket descends to scoop up raw materials at the limestone and iron ore storage area at Bethlehem Steel's Lackawanna, New York, plant.

Fig. 4-21. Skip cars are used to carry the coke, limestone, and iron ore to the top of the blast furnace. (American Iron and Steel Institute)

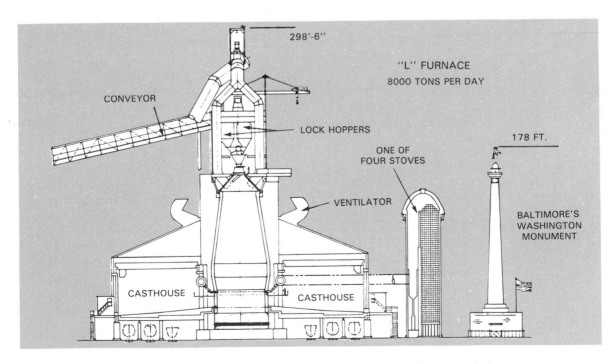

Fig. 4-22. Scale drawing of a modern blast furnace illustrates its mammoth size. This blast furnace is located at the Bethlehem Steel Corporation Sparrows Point, Maryland, plant.

Fig. 4-23. Iron and steel-making operations continue through the night at Bethlehem Steel's Bethlehem plant in eastern Pennsylvania. Blast furnaces dominate the skyline.

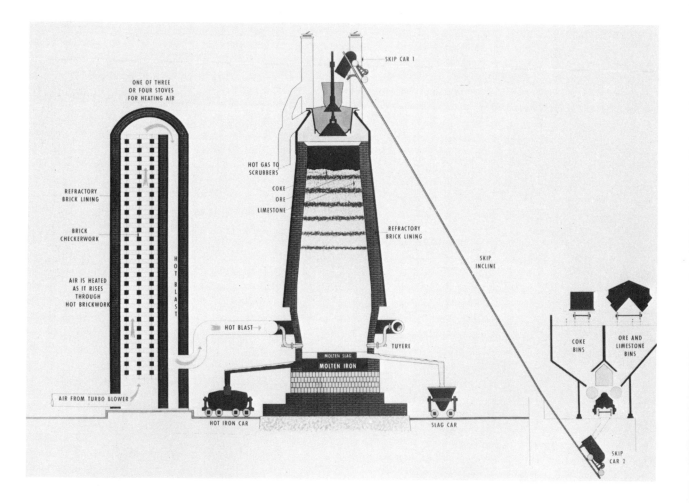

Fig. 4-24. The brickwork in the stove at the left heats the incoming air before it enters the blast furnace. Molten iron is tapped every five or six hours from this blast furnace. (Bethlehem Steel Corporation)

Inside of the stoves, there is a complex arrangement of bricks. These bricks heat up as the hot gases pass through them. Then, a short time later, the incoming air (to be used for the hot air blast in the bottom of the furnace) is reheated by the hot bricks. Study Fig. 4-24.

Modern technology in this field has come a long way. Today, technicians watch the overall blast furnace operation from a large, air conditioned CONTROL ROOM. This control room looks like a computer center with switch panels, readouts, and colored lights throughout the room. See Fig. 4-25.

TAPPING THE BLAST FURNACE

Every three to five hours, some iron (not all) is removed from the blast furnace. See Fig. 4-26.

Fig. 4-26. Every three to five hours, the blast furnaces are tapped by burning out a plug at the bottom of the furnace. (American Iron and Steel Institute)

Fig. 4-25. This control center at the Bethlehem Steel Corporation Sparrows Point, Maryland, plant is computerized. Although its blast furnace is computer-controlled, it also can be manually controlled from these consoles. The 20,000 wire terminals in this control area give an idea of the complexity of the system.

A tap hole is opened at the bottom of the furnace, and the molten iron flows along a long trough and is transferred into a BOTTLE CAR, Fig. 4-27. The bottle car looks like a submarine with railroad car wheels. It is a gigantic drum, lined on the inside with refractory (heat resistant) brick so that the iron retains its heat and stays molten. Iron enters the drum at about 2600°F (1 427°C). Then, the bottle car takes the molten iron directly to a steel-making furnace, Fig. 4-28.

The slag also is tapped off, Fig. 4-24, and collected in a large ladle. Slag has some commercial uses, especially as insulation. It is also used in brick making, as an asphalt filler, and as a spread on ice to prevent falling.

PIG IRON AND HOT METAL

The most important product from the blast furnace is the molten iron. MOLTEN IRON is commonly referred to as HOT METAL or PIG IRON.

The molten iron that comes out of the blast furnace will take one of two routes. One route involves castings as the end product, particularly cast iron castings. The other route leads to the steel-making building where some type of steel

Fig. 4-27. Molten iron flows down a trough into the bottle car that will carry the molten iron to a steel-making furnace. (American Iron and Steel Institute)

shape becomes the end product. In the first route, the molten iron is poured from the hot metal bottle car into long MOLDS which are cooled and hardened. Each mold is referred to as a PIG, which weighs about 40 lb. or more. These pigs are refined in a casting machine, then sold to foundries. Molten iron that is cooled and solidified in this manner is called PIG IRON.

Fig. 4-28. These refractory lined "bottle" or "submarine" cars carry hot metal from a blast furnace at Bethlehem Steel Corporation's Sparrows Point, Maryland, plant. These bottle cars have a capacity of 330 tons.

The term "pig iron" certainly is a strange one. Many years ago, the molten iron was poured into a long trough which had many gates or openings along its sides. The liquid metal flowed along the trough, through one of the gates, and into a sand mold shaped somewhat like a baby pig. After all the metal had cooled, it looked like an entire family of suckling pigs, all lined up at the sow's dinner table, with each mouth pressed against its gate.

In its initial state, pig iron is a rather useless material with essentially no product value. It is hard, brittle, and not very strong. However, pig iron is a vitally important ingredient in the manufacture of steel and cast iron.

Cast iron foundries (and some steel foundries) purchase the pig iron and melt it in induction furnaces or cupolas, as shown in Fig. 4-29. This metal is used to make the world's cast iron products, Fig. 4-30.

Most of the molten iron travels the second route from the blast furnace. The HOT METAL or "liquid pig iron" is carried by the hot metal bottle car to the steel-making building, Fig. 4-28.

There, is it poured into LADLES for charging into either basic oxygen or open hearth steel-making furnace, Figs. 4-31 and 4-32. The hot metal remains in the molten state and never solidifies during its journey from the spout of the blast furnace to the mouth of the basic oxygen or open hearth furnace.

STEEL-MAKING FURNACES

The three major methods of converting molten iron into steel are:
1. Open hearth furnace, Fig. 4-33.
2. Basic oxygen furnace, Fig. 4-34.
3. Electric arc furnace, Fig. 4-35.

A fourth method, the Bessemer Converter, was a common steel-making process for many years, but is used very little today.

OPEN HEARTH FURNACE

For 50 years, the primary method of manufacturing steel was the open hearth. The OPEN HEARTH FURNACE looks like a giant wash basin (hearth) that is exposed (open) to a powerful flame of fire that shoots against the contents

Fig. 4-29. This cupola furnace is used to melt and refine pig iron into cast iron.
(Central Foundry, Division of General Motors Corporation)

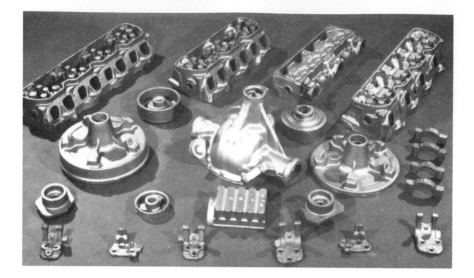

Fig. 4-30. Cast iron is used to make a variety of different parts.
(Central Foundry, Division of General Motors Corporation)

Fig. 4-31. Molten iron is charged into an open hearth furnace
from a ladle. (Bethlehem Steel Corporation)

Fig. 4-32. Molten iron from the blast furnace is charged into
a basic oxygen furnace at Bethlehem Steel Corporation's
Bethlehem, Pennsylvania, plant.

of the basin, Fig. 4-36. Scrap steel, limestone, pig iron, and some iron ore are loaded into this giant hearth and heated to about 3000°F (1 650°C). This produces a violent but spectacular boiling action. It is hard to believe that this bubbling, boiling fluid actually is steel.

An open hearth measures about 30 ft. by 90 ft. As many as 600 tons of steel may be manufactured in a single HEAT (one complete cycle that converts molten iron to steel). At most steel mills, several open hearths are located side by side for convenience in loading, tapping, and overall layout. Two floors are required. The hearth itself takes up most of the second floor, Fig. 4-33. The loading is also done on the second floor. The first floor houses the heating system.

Fig. 4-33. Molten iron is charged into two open hearth furnaces (right and far left) for converting iron into steel. (Bethlehem Steel Corporation)

Fig. 4-34. These three basic oxygen furnaces (BOF) each produce 290 tons of steel in about 50 minutes. (Bethlehem Steel Corporation)

The Open Hearth Process

The fuel most often used to heat the open hearth is natural gas mixed with air. Sometimes, liquid fuel oil, tar, or gases from the coke ovens are also used. With this process, a single heat (heating cycle) takes up to 10 hours. In recent years, the speed of the reaction has been increased by introducing an oxygen blast from overhead. The oxygen gas flows directly onto the burning ingredients in the open hearth fire.

After the burning gases pass over the molten mass in the center of the hearth, they are collected on the other side of the furnace and routed through a complex brick arrangement known as a CHECKER CHAMBER, Fig. 4-36. As these hot gases pass through the checker chamber, the bricks are heated. After 15 to 20 minutes, the burning flame reverses direction and enters the hearth from the opposite side. As incoming air

Fig. 4-35. An electric furnace is tapped of high quality steel after the steel is made to a very exact composition. (Armco Inc.)

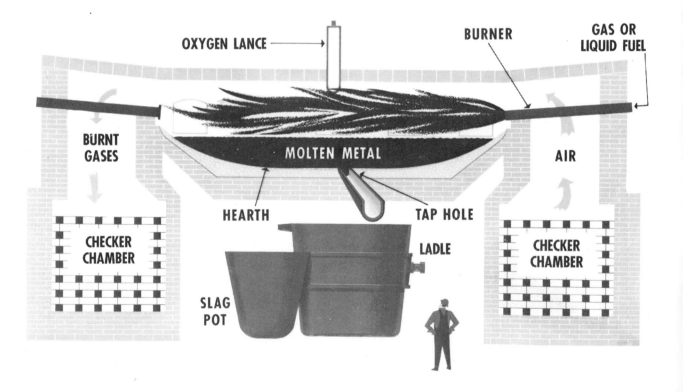

Fig. 4-36. The pool of molten metal which lies on the hearth of the open hearth furnace is exposed to the sweep of flames coming alternately from either side. The entire process of making a heat of steel may take up to 10 hours. (Bethlehem Steel Corporation)

is introduced, it passes over these same hot bricks. This heats the air and increases the efficiency of the furnace. It also reduces the amount of pollution that escapes from the furnace. Hot gases then heat bricks on the other side as they pass over a second checker chamber. This operation is reversed again and again every 15 to 20 minutes.

The molten iron is added after the other ingredients have melted into a liquid. Molten iron is poured in from a giant ladle. The molten metal pours out of the spout of the ladle into a large accepting beak at the mouth of the furnace, Fig. 4-37. At the end of the heat, when the furnace is tapped, alloys are added last.

At the end of the heat, a plug is removed from the plug hole. An explosive charge may be used to remove the plug. As the plug comes out, fresh steel flows out, along a trough, and into another giant ladle. As the steel enters the trough, sparks fly in a spectacular display that looks like fireworks on the Fourth of July.

The slag floats to the top of the ladle, where most of it is skimmed off into a SLAG THIMBLE, Fig. 4-38.

Next, the ladle of molten steel is transported to another building where it is poured into INGOT MOLDS, which give shape to the steel when it cools and solidifies.

BASIC OXYGEN FURNACE

The BASIC OXYGEN FURNACE (BOF) has replaced the open hearth process as the number one method of steel making. Capital equipment for a new basic oxygen furnace costs less, and it is a much faster way to make steel than by the open hearth. The only reason for not moving ahead of open hearth earlier was the high cost of oxygen.

The basic oxygen furnace can manufacture 300 tons of steel in less than an hour, which is five to 10 times faster than the open hearth. What

Fig. 4-37. Hot iron from the blast furnace is poured into an open hearth furnace. This molten iron will be subjected to temperatures of approximately 2900°F (1593°C) in the open hearth before being purified into steel.
(Bethlehem Steel Corporation)

Fig. 4-38. As molten steel fills the ladle at the rear of the open hearth furnace, some of the lighter slag rises to the top and flows into the adjacent slagpot. The remaining slag at the top of the ladle provides a protective blanket for the steel to retain its heat until it is ready to be poured into molds.
(Bethlehem Steel Corporation)

makes the process so successful is the oxygen blast. Massive amounts of oxygen in this blast enter the furnace at supersonic speeds and intensify the heat. The quality of the steel produced is about the same as steel made by the open hearth.

The Basic Oxygen Process

The entire basic oxygen process is rapid and colorful. It can be described in eight steps:

1. The furnace uses a huge, pear-shaped steel barrel lined with refractory material, Fig. 4-39. The barrel pivots on a shaft and tips to one side so that scrap (about 90 tons) can be poured into its mouth.
2. Immediately following the scrap charge, an overhead crane pours a ladle of molten iron (about 200 tons) from the blast furnace into the mouth of the BOF. See Fig. 4-40.
3. The OXYGEN LANCE (pipe) is lowered from above the furnace to approximately

Fig. 4-39. A basic oxygen furnace is shown in operation. (Armco Inc.)

Fig. 4-40. After scrap is charged into the basic oxygen furnace, molten iron is added. (Bethlehem Steel Corporation)

Fig. 4-41. A high pressure stream of oxygen is blown onto the molten iron and scrap at supersonic speeds through a "lance," which is inserted into the basic oxygen furnace. (American Iron and Steel Institute)

6 ft. above the surface of the metal. The lance is locked in place. The mixture is ignited, and the high pressure stream of oxygen hits the mixture at supersonic speeds, Fig. 4-41.

4. Shortly after ignition takes place, lime and fluorspar are added. These will combine with carbon and other impurities to form the slag. Temperatures of about 3000°F (1 650°C) are reached.

5. Now an intense chemical reaction takes place which keeps the molten mass churning as if in agony, Fig. 4-42. This continues uninterrupted for about 20 minutes.

6. When the chemical reaction is over, the oxygen lance is removed. The molten steel is poured into a large ladle. The slag is poured into another ladle.

7. Alloys are added after the steel is poured into the ladle.

8. The ladle of liquid steel is then taken to the ingot center, and the steel is poured into ingot molds.

Fig. 4-42. A heat shield protects members of the basic oxygen furnace crew while taking steel samples and temperature measurements from the bath of molten metal. The steel sample is sent by pneumatic tube to the basic oxygen furnace spectrometer laboratories for analysis. (Bethlehem Steel Corporation)

The entire process is computer-controlled. The computer determines the precise amounts of each charge of material to be added. It also determines cycle time for the operation.

Basic oxygen furnaces are generally installed in pairs, so one can be filled with raw material while the other is manufacturing hot steel.

ELECTRIC ARC FURNACE

The electric arc furnace differs from the basic oxygen and open hearth in three ways.
1. It uses electricity rather than gaseous fuel to produce the heat.
2. The quality of its steel can be controlled more accurately than with the other furnaces.
3. It is used primarily for special quality steel, such as stainless steel, tool steel, and high alloy steel, Fig. 4-43.

The furnace itself looks like a giant tea kettle with a spout on one end, Fig. 4-44. It is a round, tightly closed chamber. The roof pivots and swings aside so that the raw materials can be loaded into the furnace. The steel outer shell of the furnace is lined with heat resistant refractory brick. Three retractable electrodes extend up through the roof of the furnace. They may be as large as 2 ft. in diameter and 24 ft. long. These electrodes are used to ignite the metal charge. See Fig. 4-45.

The controls of the electric arc furnace can regulate the temperature more precisely because electricity is easier to control than gaseous fuel. Also no air is introduced, so there is a better control on the oxygen content.

Generally, the electric arc steel-making process is more costly than the basic oxygen or the open hearth. Therefore, it has been used only for the high quality steels. However, in recent years, the size of new electric arc furnaces has been increased, Fig. 4-45, and the electric arc process is becoming competitive in the low and medium carbon steel market.

The Electric Arc Process

The main ingredient in the electric arc furnace is steel scrap (another way in which the electric arc furnace is different from others). The scrap is the first to go in. This scrap is analyzed and examined much more carefully than the scrap that goes into any other furnace. It is separated, graded and sorted into as many as 65 different classes of steel scrap. Because of this, the composition of the finished steel can be predicted more accurately. This again emphasizes the

STEEL-MAKING FURNACE	STEEL COST	APPLICATION	RATING (BASED ON QUANTITY OF STEEL PRODUCED)
Basic Oxygen Furnace	Lowest Cost and Most Efficient	General Steel	1
Electric Arc Furnace	Slightly More Expensive	Special Steels And High Alloy Steels	2
Open Hearth Furnace	Low Cost	General Steel	3

Fig. 4-43. This chart compares the applications of the three main steel-making processes.

Fig. 4-44. Molten steel pours out of an electric furnace during a tap at Bethlehem Steel Corporation's Steelton, Pennsylvania, plant. This electric furnace has a capacity of 160 tons and can produce a heat of steel in three hours.

outstanding control of the electric arc steel-making process.

After the scrap has been introduced, the electrodes are lowered through the retractable roof into the furnace near the metal charge. See Fig. 4-45. Electricity jumps from one electrode to the metal charge, through the metal itself, then back to another electrode. Heat is developed both by the resistance of the metal to the flow of the massive amount of electricity, as well as by the heat of the arc itself.

The operation of the electric arc furnace is spectacular. As soon as the electricity is turned on, a large roar occurs, and electric arcs flash like lightning in a thunderstorm. The electric arcs strike the metal charge in the bottom of the furnace and the furnace ingredients are in turmoil. The electrodes turn white hot as more than 800,000 volt-amperes of electricity may be used. The electrical charge is so great that the cables that carry the electrical charge literally sway when the electricity is turned on, Fig. 4-46.

After the scrap has melted, a variety of other ingredients are added. Some iron ore may be used to remove carbon; or carbon may be added if the carbon content is low. Limestone is added as the fluxing agent. Alloys are dropped in at the very end. Oxygen gas is sometimes fired into the melt to speed up the heat. Mill scale and/or small amounts of molten iron may be added.

As much as 300 tons of steel may be manufactured per heat although, generally, most electric arc furnaces are smaller. About three to seven hours are required for a complete cycle. When the steel is ready, the slag is poured off first. Then the steel itself is poured into a ladle, Fig. 4-44. Eventually, it travels to the ingot center and is poured into molds. Approximately 25 percent of the steel made today is produced by electric arc furnaces.

Fig. 4-45. Heat for the electric arc furnace is made available by three large retractable electrodes which are lowered into the furnace near the charge of selected scrap iron and steel. (American Iron and Steel Institute)

Fig. 4-46. A maze of electric cables carry the electrical charge to the electrodes of an electric arc furnace. (American Iron and Steel Institute)

PROCESSING OF THE INGOT

An INGOT is a large steel casting that has cooled and solidified into a workable shape. The ingot is processed by means of three separate operations:

1. TEEMING (pouring molten steel into ingot molds).
2. STRIPPING (removing ingots from the ingot molds).
3. SOAKING (heating ingots to obtain more uniform properties in the metal).

TEEMING

When the steel arrives from any of the steel-making furnaces, it is poured into ingot molds from a large ladle, Fig. 4-47. The molten metal flows through a hole in the bottom of the ladle. A long row of ingots are lined up and poured at one time.

The shape of the ingot is normally square or rectangular in cross section, Fig. 4-48. It is tapered and has rounded corners and corrugated sides. Ingots may be anywhere from 3 ft. to 8 ft. high and 1/2 ft. to 3 ft. wide.

STRIPPING

As soon as the ingots are partially cooled, they are separated from their molds as quickly as

Fig. 4-47. Molten steel is transferred by pouring or "teeming" from the ladle into ingot molds.
(Bethlehem Steel Corporation)

Fig. 4-48. These ingot molds are used to shape the steel after it is poured from the ladle. (American Iron and Steel Institute)

possible. A stripper crane with giant tongs grips the ingot mold by lugs that are located at the top of each mold. The tongs then lift up the mold shell and separate it from the hot ingot, Fig. 4-49.

Fig. 4-49. After the steel has cooled and solidified, it is called an "ingot." The ingot is removed from the ingot mold and transferred to a soaking pit.
(American Iron and Steel Institute)

SOAKING

After stripping, the hot ingot is carried to a soaking pit which, in reality, is a furnace. It usually is below floor level, thus resembling a pit. The ingot is heated in the soaking pit for six to eight hours at about 2200°F (1 205°C) or until a uniform temperature is reached throughout the entire ingot.

If the ingots were not soaked, their outside would solidify before the inside. Under these circumstances, carbon, phosphorous, and sulfur (which tend to solidify last) would congregate around the top center portion of the ingot.

After soaking is completed, the hot ingot is removed from the soaking pit, Fig. 4-50, and carried to the rolling mill.

THE ROLLING MILL

The ROLLING MILL is the part of a steel-making complex wherein a series of large, hard rollers compress steel ingots into different shapes. An ingot that is 2 ft. thick will be "squeezed" until it is considerably thinner. Eventually, this same steel may be pressed to 1/32 in. sheets or drawn into 1/16 in. diameter wire. Some stock is compressed even smaller than this.

This "squeezing" is done by using a pair of rollers, as shown in Fig. 4-51. The ingot first passes between the two rotating rollers, which are rigidly supported in strong bearings. The space

Fig. 4-50. This ingot has been heated to an even temperature throughout and is being removed from the soaking pit. (Bethlehem Steel Corporation)

Fig. 4-51. A glowing ingot takes on a new shape as it is worked by rolls of a blooming mill. Rolling not only shapes the steel but also improves its mechanical properties. (Bethlehem Steel Corporation)

between the rollers is slightly less than the thickness of the ingot, Fig. 4-51. Since the ingot is hot and pliable, it is reduced to the thickness of the space between the rollers.

After passing between these rollers, it again moves between two rollers where the gap is slightly smaller. Here, the steel is compressed again. Then, it moves a third time to where the gap again is smaller. This squeezing continues until the steel has reached the desired thickness.

A common method of squeezing the ingot involves only one set of rollers. The ingot passes back and forth between this roller pair. After each pass, the gap between the rollers is narrowed until the desired thickness is attained. This is called a two-high reversing mill, Fig. 4-52.

There also is a three-high mill and a four-high mill, Fig. 4-53. In both cases, the metal passes between the two lower rollers and is compressed. Next, it is raised and returns back between the two uppermost rollers. The gap between the upper two rollers is slightly less than the gap between the lower two, so it is compressed further. Then, both gaps are decreased slightly and the steel repeats its journey, forward and back. After many passes, the steel is the proper thickness.

As the steel block becomes thinner, it also becomes longer. Steel from an ingot that starts out 9 ft. long may be several miles long before it comes out of the last rolling mill, Figs. 4-54 through 4-57. The steel may make as many as 50 passes before it is ready for commercial sale.

SLABS, BLOOMS, AND BILLETS

At a rolling mill, the rolling is done in two stages. The first stage is the PRIMARY ROLLING MILL. The second stage is the SECONDARY ROLLING MILL.

In the primary rolling mill, the ingot is converted into a slab (in a slabbing mill), a bloom (in a blooming mill), or a billet (in a blooming and billet mill). The slab, Fig. 4-58, receives its

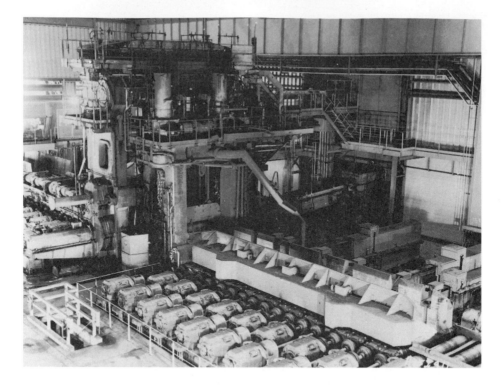

Fig. 4-52. This two-high reversing slab mill is driven by a pair of 6000 horsepower motors to convert 35-ton ingots into steel slabs. (Armco Inc.)

Fig. 4-53. This four-high finishing mill produces plates from 3/16 in. to 15 in. in thickness, up to 150 in. in width, and up to 120 ft. in length, at Bethlehem Steel Corporation's Burns Harbor, Indiana, plant.

Fig. 4-54. Steel bars travel up to 50 miles per hour through 22 stands on the 13 in. bar mill at Bethlehem Steel's Lackawanna, New York, plant. Steel bar is reduced in size in many steps through this bar mill.

Fig. 4-55. This hardened steel roll is shown in operation at a Bethlehem Steel Corporation plant. Rolls of this type are used for cold-rolling steel sheet, and strip.

Fig. 4-56. A series of rollers can be seen in this hot strip mill where the thickness of the steel is computer controlled all the way from the slag furnace at one end to the down-coilers, which are a quarter mile away at the other end. (Armco Inc.)

Fig. 4-57. This cold rolling mill has five stands with automatic gage thickness control. (Armco Inc.)

Fig. 4-58. Slabs may be further rolled into plates, which are used in ships, bridges, and machinery.
(American Iron and Steel Institute)

Fig. 4-59. Blooms are rolled into beams and other structural shapes used in the construction industry.
(American Iron and Steel Institute)

Fig. 4-60. The entire rolling operation at Bethlehem Steel Corporation's 13 in. bar mill at the Lackawanna, New York, plant is monitored from this mill pulpit, which is the mill's principal control center. The steel bar mill, nearly a half a mile long, utilizes 10 computers to guide its operations.

dimensional reduction primarily in its thickness. A typical dimension of a slab would be 10 in. thick, 6 ft. wide, and 32 ft. long. The size of the bloom, Fig. 4-59, is reduced equally in two different directions. The billet is made from a finished bloom. A billet is essentially a small bloom.

From these slabs, blooms, and billets, all other commercial shapes are made in a secondary rolling mill. These shapes include sheet, strip, bar, structural shapes, plate, pipe, tube, rod, and wire.

The entire rolling operation is monitored from a principal control center. The control center contains computers, dials, gages, and switches, resembling some complex rocket launching control room. See Figs. 4-60 and 4-61.

Fig. 4-61. Most all operations at modern steel mills are controlled in a control center. Not only furnace operation, but also rolling, hardening, and stress relieving operations are monitored and controlled for the greatest precision steel product. (Armco Inc.)

CONTINUOUS CASTING

Slabs, blooms, and billets can be made in one continuous and uninterrupted step in a process known as CONTINUOUS CASTING or STRAND CASTING. A strand casting machine used in this process bypasses the ingot operation and the primary rolling operation (where ingots are turned into slabs, blooms, and billets). This process produces a continuous length of steel.

There are three common methods of continuous casting:
1. The CURVED MOLD METHOD wherein the steel is roller straightened. See Fig. 4-62.
2. The VERTICAL CUT-OFF METHOD that uses a straight mold.
3. Combinations of the curved mold and vertical cut-off methods.

The curved mold method, which is the most popular, is done in six steps:
1. The molten steel is brought from the steel-making furnace in a ladle. A stopper at the bottom of the ladle is removed, and the metal flows into the top part, or TUNDISH, of the casting machine. See Fig. 4-63.
2. The tundish acts as a reservoir and "swallows" the entire load of steel from the ladle. At the same time, the tundish releases the molten steel in a continuous stream through an exit hole in its base. The steel flows out at a steady rate and is distributed into moving molds.
3. The hollow interior of the moving mold has inside dimensions corresponding to the width and thickness of the slab or bloom or billet that is to be formed. Lining the walls of the mold are pipes through which water flows to cool the outside surface of the metal. As this surface gradually begins to "freeze," a thin skin is formed on the material. During this freezing operation, the metal continuously moves downward, and the mold oscillates up and down in order to keep the metal from sticking.
4. The metal then moves into the rolling apron area. This area is curved and contains bending rollers and a spray apron. The secondary cooling process takes place here. The metal solidifies from the outside skin toward the center as it continues to move through the curved apron. By the time the metal reaches the bottom of the apron, it is solid throughout.
5. Next, the metal enters the STRAIGHTENER. The straightener contains rollers which reshape the slightly curved metal into a flat slab, bloom, or billet. This "leveler" then moves the long ribbon of steel onto a level table for the cutoff operation, Fig. 4-64.
6. A flame cutting torch cuts the metal to the desired, precalculated length. In order to make a straight cut, the torches move forward with the metal as they cut. A mechanism on the roller table controls this movement. After the cut is completed, the torches return to the starting position and begin a new cut. This operation is pictured in Fig. 4-65.

7. After the slab, bloom, or billet is cut to size, it usually is reheated, then taken to the secondary rolling mill for finishing.

Strand casting speed may be as fast as 15 feet per minute. The entire trip from ladle to cutoff machine takes less than one half minute.

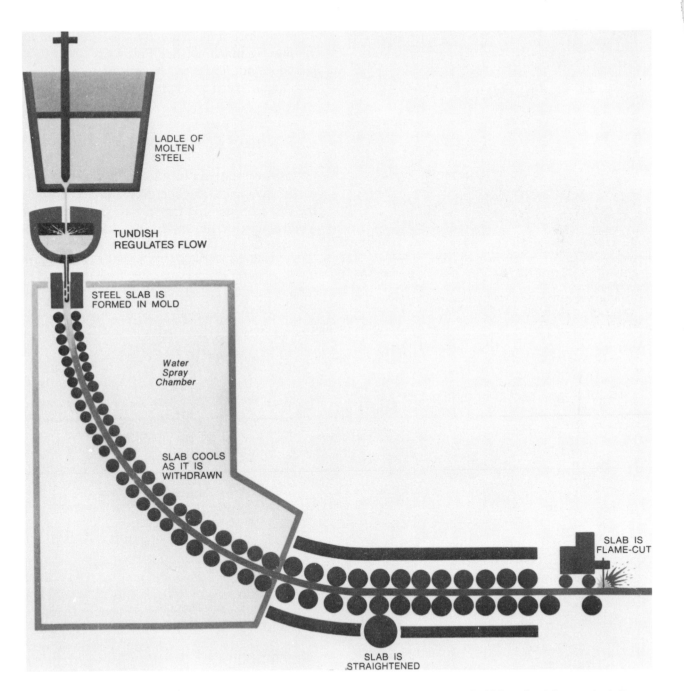

Fig. 4-62. This graphic illustration shows how the continuous slab casting process works at Bethlehem Steel Corporation's Burns Harbor, Indiana, plant. Molten steel in a ladle is teemed into a tundish, which regulates the flow of the liquid metal into a water-cooled mold. In the mold, a thin shell forms as the molten steel begins to solidify. The strand that is formed is further solidified below the mold by direct spraying with water as it is guided and supported by a curved roller apron divided into seven segments. Below the roller apron is the straightener-withdrawal section, which supports, straightens, and withdraws the cast strand as solidification is completed. The strand then moves into the slab process area, where it is cut into slabs of predetermined lengths, weighed, and stamped with identification numbers.

Fig. 4-63. An overhead crane positions the ladle of molten steel in preparation for continuous casting. The tundishes are supported on independently driven cars below the ladle arms. (Bethlehem Steel Corporation)

Vertical continuous casting machines are sometimes used but they involve less than 20 percent of the market. When continuous casting was first invented, vertical casting was regularly used. Today, in the United States, virtually all continuous casting is done by the curved mold method. Russia, reportedly still uses the vertical method. Also, in Japan, some work is being done with small, vertical machines.

MANUFACTURE OF CAST IRON

Cast iron-making furnaces convert pig iron to cast iron, Fig. 4-66. The ingredients going into the cast iron-making furnace are pig iron, scrap iron or scrap steel, limestone, and fuel.

There are thousands of cast iron manufacturers, whereas there are only a few major steel centers that cast steel. There are several reasons for this:

Fig. 4-64. Continuous casting moves a long ribbon of steel onto a level table for cutoff operations. Continuous casting can produce the longer slabs needed for huge, weld-free coils which are required in high-speed stamping and forming operations. Slabs can be cut to exact lengths. (Armco Inc.)

Fig. 4-65. Slabs are cut to required lengths by an automatic flame torch on the continuous slab caster. This two-strand caster converts 300 tons of molten steel into heavy slabs in 45 minutes. (Bethlehem Steel Corporation)

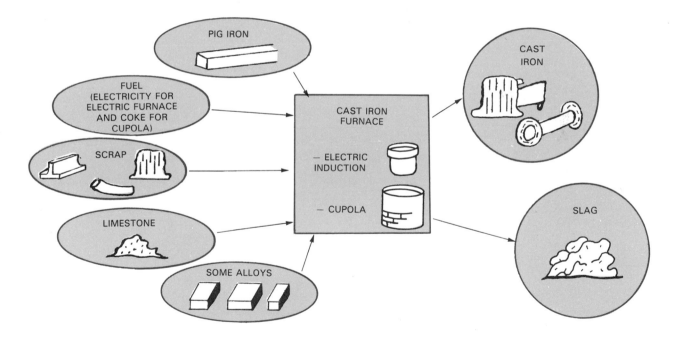

Fig. 4-66. Cast iron is made from pig iron by two basic processes: electric induction furnace or the cupola.

1. Cast iron is easier to cast. The alloying is less complicated.
2. Cast iron is slightly less expensive than steel, so more cast iron manufacturers can compete.
3. Cast iron has a superior damping ability (the ability to absorb vibration). Thousands of machine frames and other larger support parts for equipment are made of cast iron.

Therefore, most (not all) cast iron parts are large and thick, Fig. 4-67, as compared with parts that are machined out of steel.

Fig. 4-67. This diesel engine block is typical of the many large and thick parts that are made of cast iron. (Central Foundry, Division of General Motors Corporation)

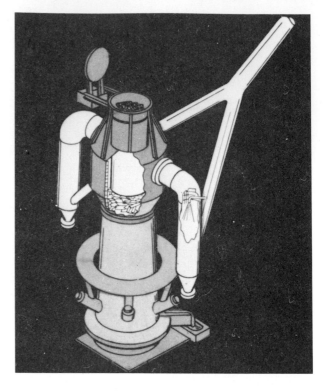

Fig. 4-68. This diagram shows pictorial representation of the gas recovery cupola. (Central Foundry, Division of General Motors Corporation)

CUPOLA

For many years, the CUPOLA served as the workhorse of the cast iron industry. Almost all cast iron was produced by the cupola prior to 1970.

The cupola is round and tall, Figs. 4-68 and 4-69. It resembles a small blast furnace. A typical cupola is two or three stories tall and measures about 5 ft. in diameter on the inside. The outer shell is made of steel. The inside is lined with refractory brick. All of the ingredients are charged into the top of the stack, Fig. 4-70. Near the bottom of the cupola, there are two discharging spouts. One spout discharges cast iron. The other spout, located a little higher on the furnace, provides the exit for the slag.

The ingredients of the cupola are pig iron, scrap iron or scrap steel, limestone, and coke.

The coke is added first. It supplies the heat. On top of the coke are layers of scrap, pig iron, limestone, more coke, more scrap, more pig iron, etc. Usually, the furnace is continuously loaded and regularly tapped, as compared with the steel-making furnace, which manufactures steel in batches. To accelerate the heat, air is charged through a WIND BOX and through TUYERES, Fig. 4-71.

The furnace normally has a sand bottom. The temperature there reaches 3700°F (2 040°C). The original charge of coke that forms the COKE BED in the bottom of the furnace is typically about 3000 lb. (1 360 kg).

The limestone helps flux out the impurities such as coke ashes, sand, and foreign matter that enter the cupola with the scrap.

Alloys are added after the cast iron is tapped from the tapping spout.

Fig. 4-69. This cupola furnace is used to melt and refine pig iron into cast iron. (Central Foundry, Division of General Motors Corporation)

INSULATED CHARGE BUCKET

TOP OF STACK

Fig. 4-70. The ingredients for a cupola are charged into the top of the stack. (Central Foundry, Division of General Motors Corporation)

ELECTRIC INDUCTION FURNACE

In the 1970s, many cupolas were removed and destroyed. The reason was fear of air pollution. The cupola burns coke. If coal dust escapes out of the stack, neighborhoods can be contaminated.

Companies whose livelihood depended on the cupola were forced to do one of three things:
1. Install very expensive antipollution cleaning equipment to their stacks.
2. Replace their cupolas with another furnace that did not burn coke, usually an electric induction furnace (also a costly route).
3. Go out of business because of insufficient funds to add the newer equipment.

The choice was a difficult one. Each of the three alternatives were taken by thousands of companies. It was a difficult time for owners of cast iron foundries, but air pollution was believed to be a realistic danger to community health.

A high percentage of the cast iron foundries took the second alternative and converted to the ELECTRIC INDUCTION FURNACE. Since the fuel for this furnace is electricity, the pollution hazard is almost eliminated. The electric

Fig. 4-71. Air is charged through tuyeres into a cupola to increase the heating process. (Central Foundry, Division of General Motors Corporation)

The electric induction furnace acts as a batch furnace (like the BOF steel-making furnace) instead of a continuous furnace (like the cupola). After loading, Fig. 4-72 and Fig. 4-73, a high frequency electric current is passed through the large copper coil. The coil acts like the primary coil of a transformer. The metal on the inside (in the crucible) acts both as the core of the transformer and as a secondary electrical coil.

As the electrical current tries to flow through the metal, the metal resists. Thus, heat is developed. The melting is very rapid and very quiet. The induced electric current in the molten metal creates a stirring action which serves to mix the metal.

Many medium-sized cast iron manufacturers installed two or more electric furnaces. One can be loaded while the other is melting and pouring cast iron. This provides a more continuous and steady supply of the molten metal. Also, having more than one furnace prevents the foundry from having to close down completely while a furnace is being cleaned or repaired.

induction furnace has other advantages for the cast iron manufacturer:
1. It can be operated economically as a small furnace, with loads of less than one ton.
2. The metal melts very rapidly in an electric induction furnace.
3. Less oxidation takes place because of this melting speed.
4. There is less chance for alloys to boil off.
Thus, the electric induction furnace became an ideal furnace for those who chose to abandon their old cupolas.

The electric induction furnace is generally round and small. The inside usually consists of a CRUCIBLE (a vessel that resists high temperatures) that is made of magnesia. The outside of the crucible is surrounded by a layer of refractory material which, in turn, has a large coil of copper tubing wound around it. The copper tubing acts as an electrical coil, and serves as the heat source.

Fig. 4-72. The raw materials needed to produce molten iron are charged into this electric induction furnace. (Central Foundry, Division of General Motors Corporation)

Fig. 4-73. These self-propelled 6000 lb. (2 720 kg) dump
buckets are used to charge the electric induction furnace.
(Central Foundry, Division of General Motors Corporation)

Aerial view shows a blast furnace (right center) at Bethlehem Steel Corporation's Sparrows Point, Maryland, plant. This blast
furnace is nearly 300 ft. high and has a rated capacity of 8000 tons of metal per day. It is computer-controlled, belt-fed, and
has the latest environmental controls, including fume evacuation from the cast house floor.

LAYOUT OF A STEEL MILL

The layout of a large steel mill is shown in Fig. 4-74. If you look closely, and study the identifying "callouts," you can recognize many of the operations that we have discussed earlier in this chapter. In the left center of the diagram, you can see the docks where the ore boats arrive with the iron ore, the coal, and the limestone. These materials are stored in piles near the docks, un-

til they are transferred to the blast furnaces.

Can you find the two blast furnace areas and the basic oxygen furnaces? Can you find the electric furnaces? A complex pattern of railroad tracks connects the shipping dock, the blast furnaces, the steel-making furnaces, the ingot mold foundry, and the various mills. Can you find the blooming mill, the plate mill, the bar and structural mills, the strip mills, the slabbing mills, and the ingot stripper buildings?

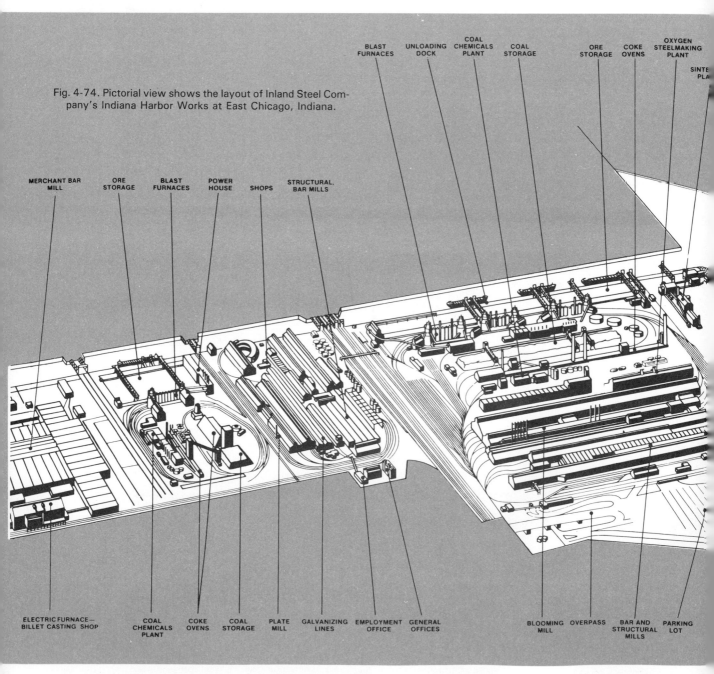

Fig. 4-74. Pictorial view shows the layout of Inland Steel Company's Indiana Harbor Works at East Chicago, Indiana.

An aerial view of another large steel plant is shown in Fig. 4-75. In the foreground, a large ore boat is docked for unloading. Just above the dock, giant piles of raw materials wait to be transported to one of the two blast furnaces in the center of the photograph. The buildings at the left house three basic oxygen steel-making furnaces and the continuous slab caster. The soaking pits, slabbing mill, and finishing facilities are in the buildings at the upper right in Fig. 4-75.

POLLUTION CONTROL

Many safeguards are employed at steel mills to protect against air pollution and water pollution. Air pollution is controlled by dust catchers, electrostatic precipitators, and wet scrubbing systems. Water is cleansed of oils and solids in settling basins, clarifiers, and treatment plants before being returned to its source or reused in closed circuit water systems.

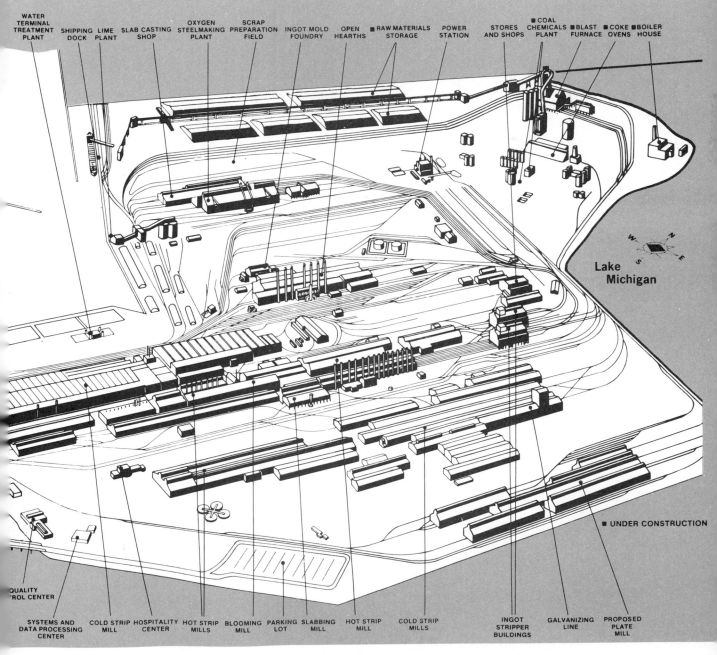

Fig. 4-75. Aerial view shows the production facilities of Bethlehem Steel's Burns Harbor, Indiana, plant, a modern, fully integrated steel plant.

TEST YOUR KNOWLEDGE

Write your answers on a separate sheet of paper. Do not write in this book.

1. What is the ingredient in a steel-making furnace that combines with impurities and forms slag?
   a. Iron ore.
   b. Pig iron.
   c. Limestone.
   d. Basic oxygen.
2. What process is most commonly used to convert iron ore to pig iron?
   a. Bessemer converter.
   b. Blast furnace.
   c. Cupola.
   d. Open hearth.
3. What is the steel-making process that is most widely used today?
   a. Bessemer converter.
   b. Basic oxygen furnace.
   c. Electric furnace.
   d. Open hearth furnace.
4. Which steel-making process is used for making the highest grades of alloy steel?
   a. Electric.
   b. Open hearth.
   c. Basic oxygen.
   d. Cupola.
5. Which steel-making process was used quite extensively at one time, but is used minimally today?
   a. Cupola.
   b. Open hearth.

c. Bessemer converter.

d. Basic oxygen process.

6. Approximately what percent carbon is in steel?

   a. 0.

   b. 1.0

   c. 3.0

   d. 8.0

7. Approximately what percent carbon is in wrought iron?

   a. 0.

   b. 1.0

   c. 3.0

   d. 8.0

8. What is the newest method for converting pig iron to cast iron?

   a. Cupola.

   b. Electric.

   c. Basic oxygen.

   d. Open hearth.

9. What is the steel-making method wherein the temperature can be controlled the most precisely?

   a. Open hearth.

   b. Bessemer.

   c. Electric.

   d. Basic oxygen.

10. What is the steel-making method wherein the furnace does not tilt on a rotary pivot?

    a. Open hearth.

    b. Bessemer.

    c. Electric.

    d. Basic oxygen.

11. Which of the following is an ingredient that goes into the blast furnace?

    a. Pig iron.

    b. Cast iron.

    c. Iron ore.

    d. Scrap steel.

12. Which steel-making furnace uses checker chambers?

    a. Open hearth.

    b. Bessemer.

    c. Electric.

    d. Basic Oxygen.

13. Which of the following is a product that comes out of a blast furnace?

    a. Cast iron.

    b. Pig iron.

    c. Iron ore.

    d. Steel.

14. Which of the following is a product that comes out of an open hearth furnace?

    a. Cast iron.

    b. Pig iron.

    c. Iron ore.

    d. Steel.

15. Which of the following is a product that comes out of a cupola?

    a. Cast iron.

    b. Pig iron.

    c. Iron ore.

    d. Steel.

16. Which of the following is a product that comes out of a basic oxygen furnace?

    a. Cast iron.

    b. Pig iron.

    c. Iron ore.

    d. Steel.

17. The temperature at the bottom of the blast furnace is approximately _____ °F.

18. The electric arc furnace is used primarily for the manufacture of:

    a. Steel.

    b. Cast iron.

    c. Pig iron.

    d. All of these.

19. After the ingot has solidified, it is reheated for several hours in a _____ pit.

20. At a rolling mill, the ingot may be rolled into either a large square known as a _____, or a thinner, flat shape known as a _____ .

21. In continuous casting, the top part of the casting machine is known as a _____, which acts as a reservoir.

22. Describe the difference in the movement of the steel in:

    a. A two-high reversing mill.

    b. A three-high mill.

5 HARDNESS

After studying this chapter, you will be able to:

- ☐ Explain what hardness is.
- ☐ Describe how the hardness of metal is found.
- ☐ Compare nine different hardness testing methods.
- ☐ Discuss how each hardness testing method works.
- ☐ Tell why hardness is so important.
- ☐ Convert one hardness value to another scale.

WHAT IS HARDNESS?

Hardness is perhaps the most important property of metals that you will encounter during your study of metallurgy. It is a difficult word to define. However, one good definition of HARDNESS is "a measure of the resistance to deformation." Another is "a measure of the resistance to penetration." Both of these definitions refer to the resistance of a metal surface to be damaged or dented or worn away or deteriorated in any way because of a force or a pressure against it.

Therefore, to invent a means of measuring hardness, one would have to create a machine with a penetrator or pointer which would try to dent or cut into a surface. A large force or weight would supply the power behind the penetrator. The size of the resulting dent or penetration in the sample would be the measure of the hardness of the material. See Fig. 5-1.

THE RELATIONSHIP OF HARDNESS TO OTHER PROPERTIES

The reason why hardness is so important in the study of metallurgy is that it relates to several

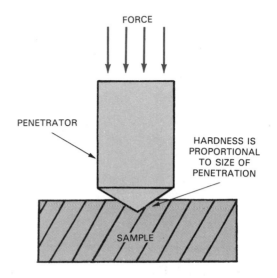

Fig. 5-1. Hardness is proportional to size of penetration.

other key properties of metal, such as strength, brittleness, and ductility. Thus, by measuring the hardness of a metal, you also are indirectly measuring the strength, the brittleness, and the ductility of that metal, Fig. 5-2.

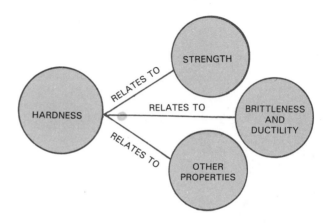

Fig. 5-2. Hardness relates to many other properties, but especially strength, brittleness, and ductility.

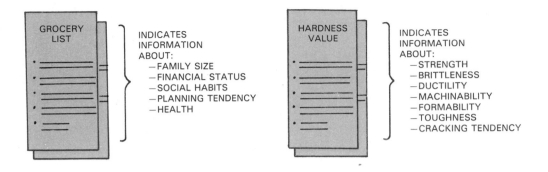

Fig. 5-3. Hardness is a gauge for other characteristics in same way that a family's grocery list indicates family traits.

Hardness, then, is similar to a family's grocery list. From the size of the list, you can estimate the size of a family. From the items of its content, you can guess the family's financial status and social habits. Similarily, hardness is a gauge for many other characteristics of a metal's family traits. See Fig. 5-3. These other traits or properties will be discussed in detail in the next chapter, Properties of Steel.

PENETRATION HARDNESS AND SCRATCH HARDNESS

The two basic categories of hardness testing methods are PENETRATION HARDNESS and SCRATCH HARDNESS.

PENETRATION HARDNESS is a very accurate measuring technique in which a precision machine is used. A penetrator on this machine is forced against the metal sample. The size of the resulting impression or dent is accurately measured, then converted to a hardness number. Penetration hardness testing is a relatively expensive and accurate method compared to scratch hardness. It is more complex and dependable, but slower. See Fig. 5-4.

A SCRATCH HARDNESS test is very fast and crude. The metal sample is scratched by the edge of a tool or object. No numerical value of hardness is calculated. The sample is called either "hard" or "soft," depending on whether or not a scratch results. See Fig. 5-4.

Most industrial companies that are concerned with hardness have some form of a penetration hardness tester.

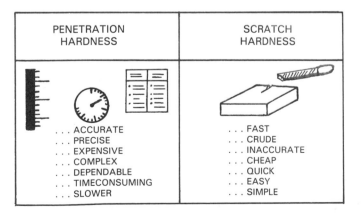

Fig. 5-4. There are two types of hardness testing methods, penetration hardness and scratch hardness.

UNITS OF HARDNESS

Time is measured in seconds or hours. Weight is measured in units of pounds or ounces or kilograms. Distance can be measured in units of feet, miles, or meters. All of these are rather obvious units of measurement. The units of hardness, however, are not so obvious, Fig. 5-5.

HARDNESS is measured in many, many different units. Some examples are: BHN, DPH, *Shore* units, *Knoop* units, R_c, 15N, and others. No one, special, popular unit is used universally as the main unit of measure. Instead, each machine tends to have its own private units. Therefore, many conversion charts are necessary to convert the units of one testing method to another hardness scale.

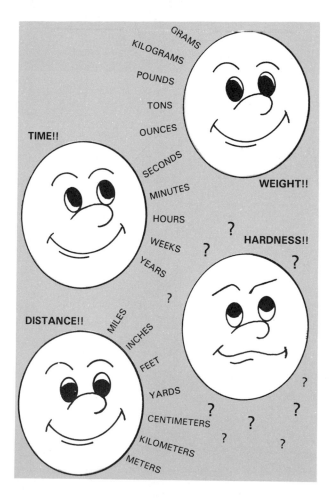

Fig. 5-5. Units of hardness are not obvious.

HARDNESS TESTING METHODS

There are many different hardness testing methods and many different hardness testing machines. Nine of the most common ones will be discussed in detail in this chapter. Each method has advantages and disadvantages over each of the other methods. Therefore, determination of which method to use or which machine to buy depends on the individual application.

The nine methods of testing the hardness of a metal surface are:
1. Brinell Hardness Testing Method
2. Vickers Hardness Testing Method
3. Knoop Hardness Testing Method
4. Rockwell Hardness Testing Method
5. Rockwell Superficial Hardness Testing Method
6. Shore Scleroscope Hardness Testing Method
7. Sonodur Hardness Testing Method
8. Moh Hardness Testing Method
9. File Hardness Testing Method

BRINELL HARDNESS TESTING METHOD

BRINELL hardness testers are shown in Figs. 5-6 and 5-7. The testing method is illustrated in Fig. 5-8. Brinell is one of the oldest methods of hardness testing.

Process Description

Four separate steps are involved in a Brinell hardness test:
1. The metal sample to be tested is placed on the anvil of the machine.
2. The hardened steel penetrator (round ball) is slowly brought into contact with the test sample, Fig. 5-9, either automatically or by manual operation. This contact pressure between the ball penetrator and the test sample becomes larger and larger until a force of 3000 kilograms is reached. Since the ball penetrator is harder than the sample, a round dent is impressed onto the sample. See Fig. 5-10.
3. After the ball has made the round indenture in the sample, it is released and the sample

Fig. 5-6. Metal sample is tested on Brinell hardness tester. (Acco Industries, Wilson Instrument Division)

Fig. 5-7. Another Brinell hardness tester is pictured. (Engineering and Scientific Equipment Ltd.)

is removed.

4. A small microscope with a calibrated lens is brought into contact with the sample, Fig. 5-11. The diameter of the dent is measured in millimeters. The millimeter reading from the small microscope is converted to a hardness value. This conversion can be done by means of the formula shown in Fig. 5-12 or, more commonly, by using a conversion chart.

The conversion chart shown in Fig. 5-13 has two columns. The left column lists the diameter of the indenture in millimeters. The right column is a BHN column which gives the hardness value in units of BHN (Brinell Hardness Number). These BHN values in charts were prepared by people who used the formula shown in Fig. 5-12.

Details of Machine

The tungsten carbide ball penetrator is approx-

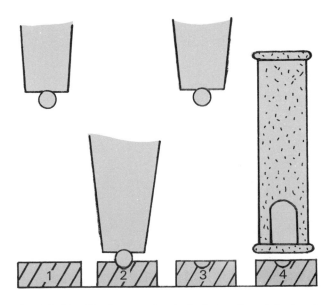

Fig. 5-8. Brinell hardness testing method: 1 — Sample is placed on anvil of machine. 2 — Penetrator contacts and indents sample. 3 — Penetrator is released. 4 — Microscope with calibrated lens is used to measure diameter of dent.

Fig. 5-9. In Brinell hardness testing method, test sample is placed on anvil. Then, a hardened steel penetrator contacts surface of test sample.
(Acco Industries, Wilson Instrument Division)

imately 10 millimeters in diameter. Generally, the force (load) used in the test is 3000 kilograms. Sometimes a second BHN scale is used in conjunction with a 500 kilogram force. This scale is used primarily for softer and thinner samples. In making the tests, loads should be applied against the sample for a minimum of 15 seconds. The sample should be relatively smooth, flat, clean, and horizontal.

The round ball makes a relatively large impression on the sample, compared to other hardness testing methods. Many of the other methods use a sharp pointed penetrator which will not deform the sample as much. Because of the greater area of penetration, Brinell samples are generally scrapped after being tested. Also, because of the large indenture, a Brinell hardness tester cannot be used to measure the hardness of very thin samples.

The Brinell method generally is restricted to softer steels or other softer metals. Too much force is required on the tungsten carbide ball to make a measurable dent on a very hard surface. However, Brinell is a very accurate hardness measuring method particularly for soft materials.

Typical BHN values are shown in Fig. 5-14.

Fig. 5-10. After Brinell hardness test is completed and penetrator is released, a round dent remains on sample. (Acco Industries, Wilson Instrument Division)

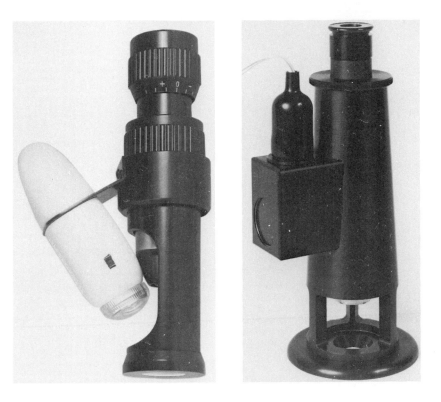

Fig. 5-11. Brinell microscope is used in conjunction with hardness tester.
(Left. Acco Industries, Wilson Instrument Division.
Right. Engineering and Scientific Equipment Ltd.)

$$BHN = \frac{\text{LOAD (kg)}}{\text{SURFACE AREA (mm}^2\text{)}}$$

$$BHN = \frac{F}{\frac{\pi D}{2}\ (D - \sqrt{D^2 - d^2})}$$

BHN = Brinell Hardness Number
F = Force or Load in Kilograms
D = Diameter of Ball Penetrator in Millimeters
d = Diameter of Indentation in Millimeters

Fig. 5-12. Formula can be used to calculate Brinell hardness
number after diameter of dent is read in millimeters.

BRINELL			
10 m/m Ball		**3000 kgm Load**	
Diam. of Ball Impression in m/m	Hardness Number	Diam. of Ball Impression in m/m	Hardness Number
2.25	745	3.40	321
2.30	710	3.45	311
2.30	710	3.50	302
2.35	682	3.55	293
2.35	682	3.60	285
2.40	653	3.65	277
2.45	627	3.70	269
		3.75	262
2.55	578	3.80	255
2.55	578	3.80	255
2.60	555	3.85	248
2.60	555	3.90	241
2.65	534	3.95	235
2.70	514	4.00	229
		4.05	223
2.75	495	4.10	217
2.75	495	4.15	212
2.80	477	4.25	203
2.85	461	4.35	192
2.90	444	4.40	187
2.90	444	4.50	179
2.95	432	4.60	170
3.00	415	4.65	166
3.00	415	4.80	156
3.05	401	4.80	156
3.10	388	4.90	149
3.10	388	5.00	143
3.15	375	5.10	137
3.20	363	5.20	131
3.25	352	5.30	126
3.30	341	5.40	121
3.35	331	5.50	116
3.35	331	5.60	112

Fig. 5-13. Chart can be used to determine Brinell hardness number after diameter of dent is read in millimeters. (Teledyne Vasco)

VICKERS HARDNESS TESTING METHOD

VICKERS hardness testing machines are shown in Figs. 5-15 and 5-18. The operation of the Vickers Hardness Tester, Fig. 5-16, is similar to the operation of the Brinell Tester. There are three main differences:

1. The penetrator has a different shape.
2. The load (force) is less.
3. The units that it reads in are different. See Fig. 5-17.

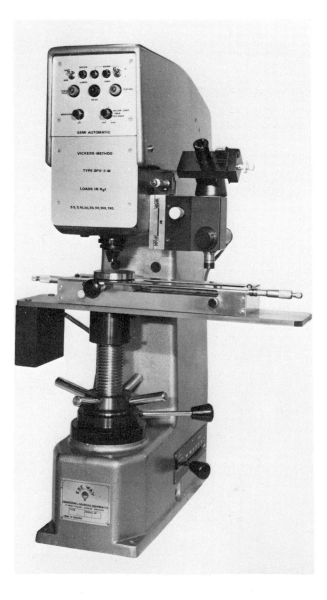

Fig. 5-15. Vickers hardness tester is pictured. (Engineering and Scientific Equipment Ltd.)

TYPICAL BHN VALUES	
Cold Rolled Steel (unquenched)	150 BHN
Quenched Steel	600 BHN
Stainless Steel (unquenched)	150 BHN
Cast Iron	200 BHN
Wrought Iron	100 BHN
Aluminum	100 BHN
Annealed Copper	45 BHN
Brass	120 BHN
Magnesium	60 BHN

Fig. 5-14. Brinell hardness values vary for different types of metal.

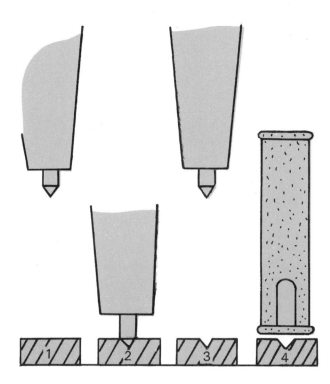

Fig. 5-16. Diagram illustrates operation of Vickers hardness tester. See steps 1, 2, 3, and 4 in Fig. 5-8.

COMPARISON OF BRINELL & VICKERS HARDNESS TESTING		
	Brinell	Vickers
Penetrator	10mm Diameter Ball	Square Based Diamond
Load	500 kg or 3000 kg	50 kg
Units	BHN	DPH

Fig. 5-17. Chart compares Brinell and Vickers hardness testing.

Process Description

Four separate steps are involved in a Vickers hardness test:

1. The sample to be tested is placed on the anvil of the tester, below a hardened steel penetrator with a diamond point, Fig. 5-16.
2. This "square based diamond penetrator" is

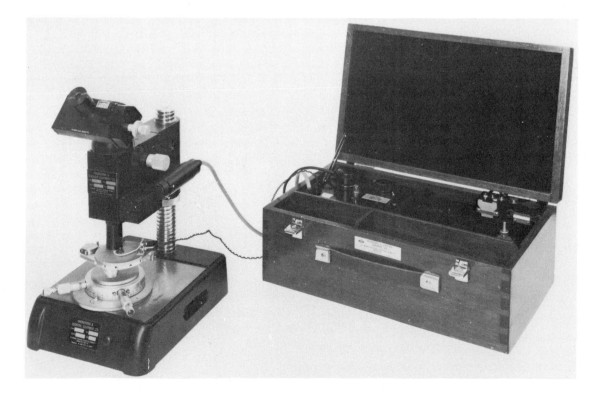

Fig. 5-18. Vickers microscope hardness tester is compact and portable. (Engineering and Scientific Equipment Ltd.)

slowly brought into contact with the sample. The contact pressure between the penetrator and sample is increased until 50 kilograms is reached.

3. The penetrator is retracted and the sample shows a small, pyramidal shaped hole, Figs. 5-19 and 5-20.

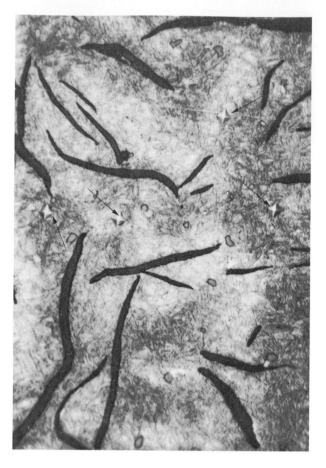

Fig. 5-20. Four Vickers "dents" can be seen in this magnified view. This gray iron is much harder than material shown in Fig. 5-19. Therefore diamond shaped dents are much smaller. (Iron Castings Society)

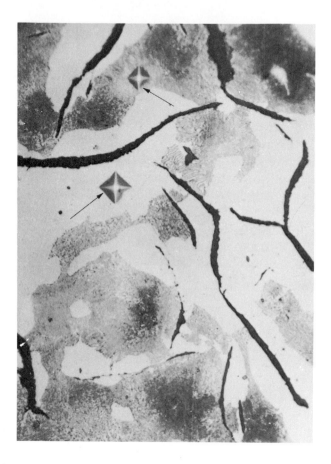

Fig. 5-19. Magnified view shows cast iron part that was tested twice on Vickers Hardness Tester. Two dents look like diamonds or pyramids. Dent in white area is larger than dent in grayish area because iron structure is harder in grayish area. Therefore, gray area is able to resist denting better than white area. (Iron Castings Society)

4. The sample is removed and a small microscope is used to measure the diagonal of the dent. (Similar to Step 2 of the Brinell method.) The length of the diagonal is converted to a DPH value by use of a formula, Fig. 5-21, or a table, Fig. 5-22. DPH is Vicker's unit of measure. It stands for "Diamond Pyramidal Hardness."

$$DPH = \frac{LOAD\ (kg)}{SURFACE\ AREA\ (mm^2)}$$

$$DPH = \frac{1.85\ F}{d^2}$$

DPH = Vickers Diamond Pyramidal Hardness
F = Force or Load in Kilograms
d = Diagonal Length of Identation in Millimeters

Fig. 5-21. This formula is used to determine DPH number for Vickers hardness test.

VICKERS HARDNESS CONVERSION	
DIAGONAL LENGTH OF IMPRESSION IN "mm"	DPH (50 kg)
.30	1030
.35	757
.40	579
.45	458
.50	371
.55	306
.60	258
.65	219
.70	189
.75	165
.80	145
.85	128

Fig. 5-22. This table is used to determine DPH number for Vickers hardness test.

Details of Machine

The angle of the penetrator is approximately 136 degrees, Fig. 5-23. The load most commonly applied is 50 kilograms. Some Vickers Testers occasionally use either 5, 10, 20, 30, or 100 kilograms. The load is held on the sample for about 30 seconds. The surface should be smooth, flat, clean, and horizontal before testing begins.

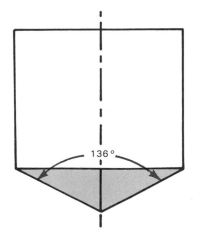

Fig. 5-23. Angle of penetrator in Vickers hardness test is approximately 136 degrees.

Note that the 50 kilogram load is considerably less than the 3000 kilogram load that was used with the Brinell test. Therefore, the sample is not damaged as severely as it is in the Brinell hardness test.

Advantages

Advantages of the Vickers method over the Brinell method are:
1. It can be used on harder materials because a pointed penetrator can probe into a hard surface more easily than a ball penetrator can.
2. It can be used on smaller areas.
3. A smaller load is required.

KNOOP HARDNESS TESTING METHOD

The KNOOP method of hardness testing is frequently called the "microhardness method." See Fig. 5-24 and 5-25. It is used for testing very small surface areas. Often it is used to test areas smaller than the size of a crystal or grain. To do this, a sharp pointed penetrator is necessary and a very small load is used.

Fig. 5-24. Microhardness tester is designed to test small surface areas. (Acco Industries, Wilson Instrument Division)

91

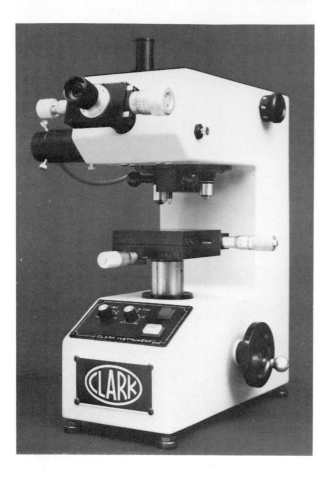

Fig. 5-25. Another microhardness tester is pictured.
(Clark Instrument, Inc.)

The penetrator does not have a square base like the Vickers method. It has a diamond cross section known as an "elongated pyramid." The ratio of the diagonals is 7 to 1. See Fig. 5-26.

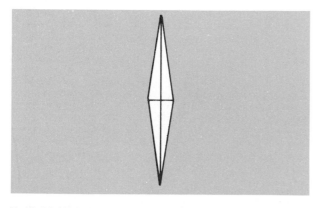

Fig. 5-26. With Knoop or microhardness test method, dent from penetrator has shape of elongated pyramid.

Process Description

The steps involved in making tests by the Knoop method are identical to the Vickers and Brinell methods. A load is applied. An indenture is made. The indenture is measured. A formula or chart is used to convert to a Knoop hardness value. The units of the Knoop test are not BHN nor DPH. The hardness units are simply called "units *Knoop*."

A load of less than four kilograms is used on the Knoop tester. A load of merely 25 grams can be used for extremely small areas.

The Knoop method finds its main application in the research laboratory rather than in the manufacturing plant. Manufacturing generally requires tests to determine the average hardness of a larger surface area.

Advantages

Advantages of the Knoop method over the other two hardness testing methods include:
1. Essentially no damage to the specimen.
2. The ability to test very thin materials.
3. The ability to test very small surface areas.

Like the other two methods, the surface of the specimen should be smooth, flat, clean, and horizontal before testing.

ROCKWELL HARDNESS TESTING METHOD

The ROCKWELL testing method and the Rockwell Hardness Testing Machine are the most widely used of all hardness testing machines. Because a preliminary load, called the "minor load," is applied to the sample before the hardness test is taken, Rockwell eliminates the bad effects of small surface imperfections. Thus, Rockwell values are considered to be very accurate readings. A variety of Rockwell hardness testers is shown in Fig. 5-27.

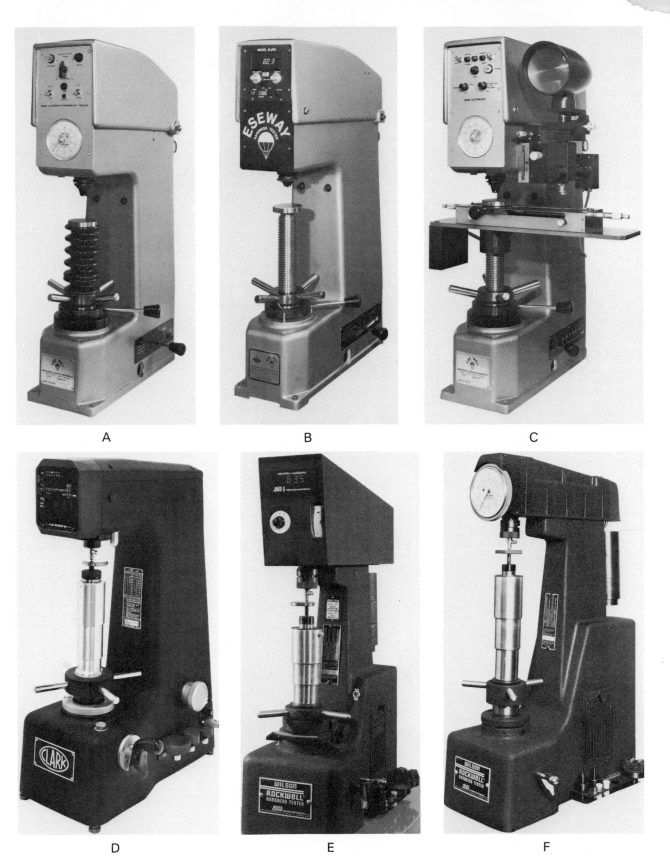

Fig. 5-27. Rockwell hardness testers vary in design and readout method, but function is same.
(A,B,C — Engineering and Scientific Equipment Ltd. D — Clark Instrument, Inc. E,F — Acco Industries, Wilson Instrument Division)

Process Description

The Rockwell test is done in two loading steps: one step is preliminary; the other is the actual hardness test.

Three separate steps are involved in a Rockwell hardness test:

1. MINOR LOAD. The sample is placed on the anvil. The anvil is raised manually until the sample contacts the penetrator. Then, the sample is raised slightly higher until a minor load of about 10 kilograms is applied, Fig. 5-28. This minor load causes the penetrator to dig slightly into the sample. Then, step 2, the actual hardness test, is begun.

Note: because this minor load is applied first, the hardness value is measured slightly below the surface of the sample, instead of on the outside surface of the metal. This, therefore, almost eliminates the bad effects of surface scale, surface roughness and lack of flatness, smoothness, and cleanliness.

Fig. 5-28. In Rockwell hardness test, minor load of about 10 kilograms is first applied. (Clark Instrument, Inc.)

2. MAJOR LOAD. After the minor load of 10 kilograms is applied, the major load (60 kg, 100 kg, 150 kg) is applied by actuating a handle or lever on the front of the machine, Fig. 5-29. As this major load is applied, the penetrator moves deeper into the sample, Fig. 5-30. A schematic representation of the major and minor loading is shown in Fig. 5-31.

3. The hardness value is read directly off a scale on the machine. See Fig. 5-32. No intermediate microscope or human approximations need to be made. The scale will read in a Rockwell hardness value, which is based on the depth of penetration. The machine automatically converts this depth reading to a Rockwell hardness value.

Advantages

You can see that the Rockwell method has two key advantages:

1. Because of the minor load, surface imperfec-

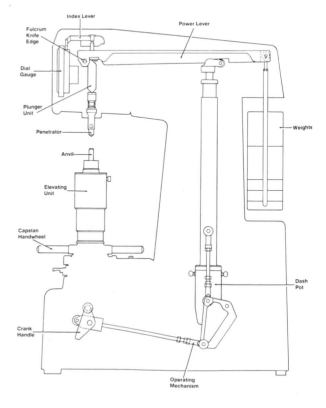

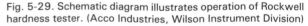

Fig. 5-29. Schematic diagram illustrates operation of Rockwell hardness tester. (Acco Industries, Wilson Instrument Division)

tions have little effect.

2. Because the hardness value can be read directly off a scale, human error is almost eliminated.

Fig. 5-30. As major load is applied, penetrator moves deeper into sample. (Acco Industries, Wilson Instrument Division)

Details of Machine

Three different penetrators may be used on a Rockwell machine, and three different loads are commonly applied. Therefore, there are nine possible combinations of penetrator and load with the Rockwell system, Fig. 5-33. The other methods which we have discussed, generally, used only one scale. Brinell used the BHN scale. Vickers used the DPH scale. Knoop read the hardness in *Knoop* units. Rockwell has nine scales: R_A, R_B, R_C, R_D, R_E, R_F, R_G, R_H, and R_K, Fig. 5-34.

The three Rockwell penetrators are:
1. A 1/8 in. diameter tungsten carbide ball.
2. A 1/16 in. diameter tungsten carbide ball.
3. A diamond point penetrator.

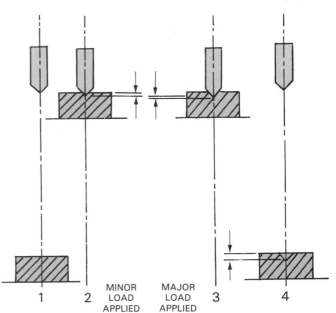

Fig. 5-31. Rockwell hardness test: 1 — Sample is placed on anvil. 2 — Sample contacts penetrator and minor load is applied. 3 — Major load is applied and scale is read. 4 — Sample is lowered.

The diamond point penetrator has a sharp point and is conically shaped, Fig. 5-35. The 1/16 in. diameter ball penetrator is shown in Fig. 5-36. The loads used on the Rockwell tester are 60 kg, 100 kg, or 150 kg.

Fig. 5-34 shows the relationship between the three penetrators, the three loads, and the nine Rockwell scales. Which scale would be used for the hardest materials? Hard materials require the largest kilogram load and the sharpest penetrator. Therefore, the "C" scale is used for hardest materials.

Can you understand why the "H" scale is used for the softest materials? This scale is so soft that it normally would not be used for any metals.

For testing steel, the two scales that you will use most often will be R_C and R_B. R_C is used to test hard steels. R_B is used to test the softer low carbon steels. R_B is also used to test aluminum and other softer nonferrous materials. The dial on most Rockwell machines has both a "B" and "C" scale. The hardest value that steel can attain is about 70 R_C.

Fig. 5-32. Rockwell hardness tester dial reads directly in Rockwell hardness value. (A—Acco Industries, Wilson Instrument Division B—Engineering and Scientific Equipment Ltd. C—Clark Instrument, Inc.)

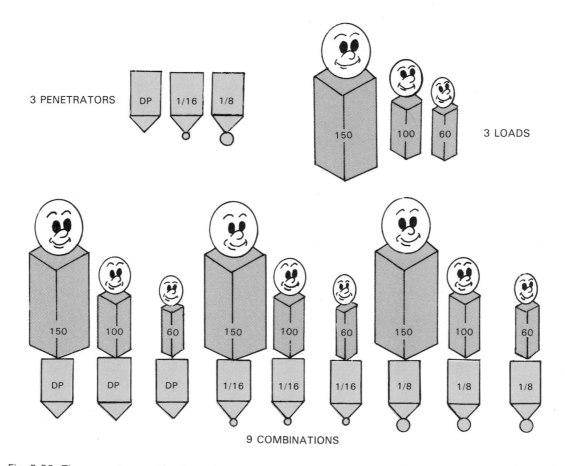

Fig. 5-33. There are nine combinations of penetrators and loads possible with Rockwell hardness testing method.

ROCKWELL SCALE	PENETRATOR	LOAD
C	Diamond Point	150 kg
D	Diamond Point	100 kg
A	Diamond Point	60 kg
G	1/16 Ball	150 kg
B	1/16 Ball	100 kg
F	1/16 Ball	60 kg
K	1/8 Ball	150 kg
E	1/8 Ball	100 kg
H	1/8 Ball	60 kg

Fig. 5-34. Table relates nine Rockwell scales to their penetrators and loads.

Fig. 5-35. Diamond point penetrator that is used on Rockwell hardness tester. (Clark Instrument, Inc.)

Fig. 5-36. A 1/16 in. diameter ball penetrator is also used on Rockwell hardness tester.
(Acco Industries, Wilson Instrument Division)

ROCKWELL SUPERFICIAL HARDNESS TESTING METHOD

The ROCKWELL SUPERFICIAL hardness tester is shown at A in Fig. 5-37. The method of testing with this machine is identical to that of the basic Rockwell hardness tester. The difference is that the Rockwell Superficial tester tests the hardness closer to the outside surface of the metal than the Rockwell tester does. See Fig. 5-38.

Views B and C in Fig. 5-37 show hardness testing machines that can be altered to test either Rockwell hardness or Rockwell Superficial hardness.

Process Description

In order to test the hardness closer to the surface, a smaller minor load and smaller major load are used. The three common loads used on the superficial tester are 15 kg, 30 kg, and 45 kg. The same diamond point penetrator and 1/16 in. ball used in the Rockwell test are also used on the Rockwell Superficial tester.

Details of Machine

Since there are two choices of penetrators and three choices of kg loadings, there are six total combinations available for use on the superficial tester. Thus, the superficial tester employs six common scales: 15N, 30N, 45N, 15T, 30T, 45T. Fig. 5-39 shows the relationship of these scales to their penetrators and kg loadings.

For example, 55-15N would mean that the material has a hardness value of 55 units. The sample was measured on a Rockwell Superficial test machine where 15 kg and a diamond point penetrator were used to run the test. A reading of 75-30T would mean that the material hardness is 75 units and the 1/16 in. ball and 30 kg load were used.

The penetrator-load combination on a Rockwell Superficial tester for the hardest steels would be the one employed for the 45N scale. This combination makes use of the diamond point penetrator and the 45 kg load.

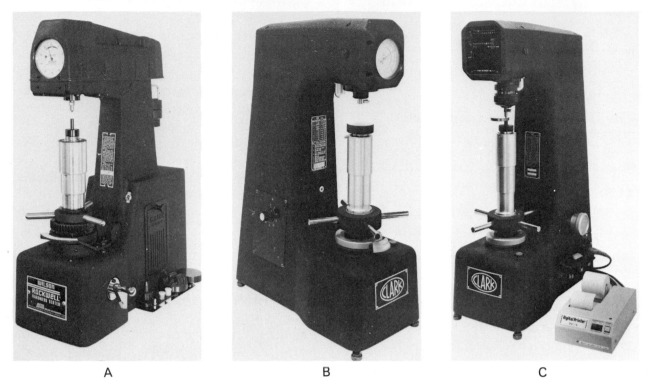

A   B   C

Fig. 5-37. Rockwell Superficial hardness tester (A) and Combination Rockwell-Rockwell Superficial hardness testers (B,C) are pictured. (A—Acco Industries, Wilson Instrument Division. B,C—Clark Instrument, Inc.)

Advantages

Advantages of the Superficial tester are:
1. Thin materials can be tested.
2. Hardnesses close to the surface can be tested.
3. Case-hardened surfaces can be tested. (Case hardening will be discussed in detail in Chapter 15.)

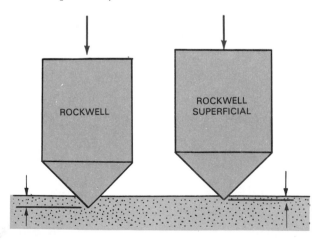

Fig. 5-38. Indenter in Rockwell Superficial hardness tester does not penetrate as deeply as penetrator in Rockwell hardness test.

Many companies use Rockwell Superficial test machines for all hardness tests. Even though their special purpose is testing hardnesses that are close to the outside surface of a metal, they are dependable for most all manufacturing applications.

ROCKWELL SUPERFICIAL SCALE	PENETRATOR	LOAD
45N	Diamond Point	45 kg
30N	Diamond Point	30 kg
15N	Diamond Point	15 kg
45T	1/16 Ball	45 kg
30T	1/16 Ball	30 kg
15T	1/16 Ball	15 kg

Fig. 5-39. Table relates six Rockwell Superficial scales to their penetrators and loads.

SHORE SCLEROSCOPE
HARDNESS TESTING METHOD

The SHORE SCLEROSCOPE method is entirely different from any other method discussed thus far. The surface is not penetrated by a diamond point or a ball. The hardness is proportional to the bounce of a ball or hammer. Two Shore Scleroscope models are shown in Figs. 5-40 and 5-41.

Process Description

The sample to be tested rests on an anvil. A small metal ball or hammer drops from a predetermined height of 10 in. The hammer strikes the test sample and rebounds. This hardness testing method assumes that the higher the hammer rebounds, the harder the material is.

The height of the first bounce is monitored. This height is converted into a hardness value with units of *"Shore."* For example, if the hammer bounces 6 1/4 in. after it strikes the sample, the hardness is 100 units *Shore*. If the hammer bounces 3 1/8 in. high, the hardness value is 50. If the hammer would strike the sample and

Fig. 5-40. Shore Scleroscope tests hardness by measuring height of hammer bounce.
(The Shore Instrument & Mfg. Company, Inc.)

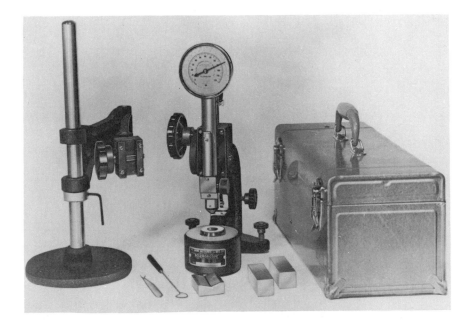

Fig. 5-41. Shore Scleroscope, clamping stand, and carrying case are displayed. One advantage of this test machine is its portability. (The Shore Instrument & Mfg. Company, Inc.)

not bounce, the hardness of the material would be rated 0 units *Shore*, Fig. 5-42.

This method may sound very unorthodox. However, the correlation between *Shore* values and Rockwell and Brinell values is very close.

Advantages

Advantages of the Shore Scleroscope method are:
1. The impression made is negligible since nothing actually penetrates the surface.
2. The machine is small and portable, Fig. 5-41. It can be carried about the factory. A conventional Rockwell or Brinell tester is too large to conveniently maneuver down the aisles of a manufacturing plant.

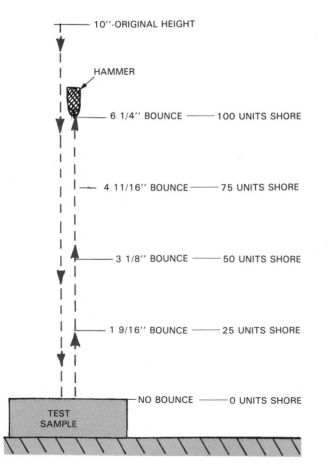

Fig. 5-42. Chart relates height of hammer bounce to units of hardness in a Shore Scleroscope.

Details of Machine

The hammer weighs 40 grains. This is less than .001 pound; less than 3 grams.

On most Shore machines, the scale has a follower which follows the first bounce. Therefore, the reading can be read directly off a dial, with little human reading error.

The need of having a surface that is smooth, flat, clean, and horizontal is of greater importance with the Shore Scleroscope method than with most other hardness testing methods. An imperfection in the surface noticeably affects the movement of the hammer.

SONODUR HARDNESS TESTING METHOD

SONODUR employs a very unusual method of hardness testing. It makes use of the fact that the hardness of a material can affect the natural resonant frequency of a piece of metal.

Process Description

In making the Sonodur hardness test, a diamond-tipped magnetostrictive rod (0.75 millimeters in diameter) is pressed against the sample to be tested. An electrical coil is used to vibrate the rod, and the frequency at which it vibrates the greatest is determined in the electronic portion of the machine. This frequency is known as the RESONANT FREQUENCY. The harder the material is, the higher this resonant frequency will be.

The resonant frequency value is then converted to a hardness number. The Sonodur machine reads in BHN.

Advantages

The Sonodur machine is small and portable. It gives a very quick response time and does not damage the specimen. The Sonodur hardness testing method is newer than most other methods, and is considered very accurate.

MOH HARDNESS TESTING METHOD

The MOH Hardness Scale was probably the first hardness testing method ever invented.

In ancient times, the philosophers of the day decided that they had to come up with a hardness scale. They selected ten stones of varying hardnesses. The softest stone was given a hardness value of one. The hardest stone was given a hardness value of ten. The other eight were given values in between one and ten. These ten stones are listed in the table in Fig. 5-43.

MOH SCALE OF HARDNESS
1 — Talc
2 — Gypsum
3 — Calcite
4 — Fluorspar
5 — Apatite
6 — Orthoclase
7 — Quartz
8 — Topaz
9 — Sapphire
10 — Diamond

Fig. 5-43. Ten stones used in Moh hardness scale.

Here is how the testing was done:

The material sample was struck with one of the stones. If it scratched, that indicated that the unknown material was softer than the stone. Therefore, a softer stone was used to try to scratch the sample. This procedure was continued until a stone was found that did not scratch the sample. For example, suppose that topaz and quartz and orthoclase scratched your sample but apatite and fluorspar did not. Then, the hardness of your sample would be between 5 and 6 units Moh. This certainly was not an accurate method but it was the best that they had at the time. Today, this method is no longer applied, except for a minimal use in the field of geology.

FILE HARDNESS TESTING METHOD

The FILE HARDNESS test is fast, simple, and convenient — but inaccurate. However, it does give a quick answer, so it is used extensively in industry today. See Fig. 5-44.

To do the file hardness test, the inspector takes a file in one hand and the test material in the other. An attempt is made to scratch the test material by scraping it once with the edge of the file. If the material does not scratch, it is said to be "file hard." If it does scratch, it is said to be "not file hard."

This may sound very crude and useless as a hardness testing method, but often in industry, one does not care about a specific numerical hardness value. Often an inspector merely wants to check if some parts have received a hardness treatment or not. With one quick stroke of a file, they have their answer.

Fig. 5-44. File hardness testing method is used extensively in industry because it gives quick results.

This test method is not intended to give an accurate hardness value, like BHN or R_C or DPH. However, many inspectors who have performed the file hardness test for years are very skilled at it and may argue that they can tell the numerical hardness value within a few points. One is inclined to believe these experienced people.

Since the file hardness test is dependent on the sensitivity of the hand of the inspector and on the sharpness of the file, most companies will rely on only one or two experienced people in their factory to run the test. Then, the results can be compared dependably. It is also important to use a relatively new file so that the hardness of the file does not cause a variation in the test.

COMPARISON OF HARDNESS TESTING METHODS

All nine of the hardness testing methods are summarized in Fig. 5-45. This chart will give you a quick check of the relative advantages and disadvantages of each technique.

CONVERSION SCALES

Since hardness can be measured in so many different units, using different machines and scales, there must be a simple way to convert one hardness scale value to another scale value. If company A has a Rockwell tester and company B has a Brinell hardness test machine, and they are working together in manufacturing the same parts, they must be able to communicate with each other.

The answer lies in the use of CONVERSION CHARTS such as Figs. 5-46 and 5-47. With charts of this kind, hardness values can be converted to other scales that measure in the same hardness range. For example, a 52 R_C hardness reading that was measured on a Rockwell "C"

METHOD	BASIS	PENETRATOR	LOAD	READING SYMBOL
Rockwell	Depth of Penetration	Diamond Point or 1/16-1/8 Ball	60-100-150 kg	R_C etc
Rockwell Superficial	Depth of Penetration	Diamond Point or 1/16-1/8 Ball	15-30-45 kg	15N 30T etc
Brinell	Area of Penetration	10 mm Ball	500-3000 kg	BHN
File	Appearance of Scratch	File	Manual	None
Shore	Height of Bounce	40 Grain Weight	Gravity	Units Shore
Vickers	Area of Penetration	Pyramidal Diamond	5 to 120 kg	DPH
Knoop	Area of Penetration	Pyramidal Diamond	25-3600 Grams	Units Knoop
Sonodur	Frequency of Vibration	Magneto-strictive Rod	N.A.	BHN
Moh	Appearance of Scratch	10 Stones	Manual	Units Moh

Fig. 5-45. Chart compares hardness testing methods.

| ROCKWELL | | | | SUPERFICIAL ROCKWELL | | | BRINELL 10 m/m Ball 3000 kgm Load | | Vickers of Firth Diamond Hardness Number | Scleroscope | Tensile Strength Equivalent 1000 Lb. Sq. In. |
| Diamond Brale | | | 1 16" Ball | "N" Brale Penetrator | | | | | | | |
150 kgm C Scale	60 kgm A Scale	100 kgm D Scale	100 kgm B Scale	15 kg Load 15 N	30 kg Load 30 N	45 kg Load 45 N	Diam. of Ball Impression in m/m	Hardness Number			
68	86	...	...	93	84	75	...	...	940	97	...
68	85	...	...	93	84	75	...	...	920	96	...
67	85	...	...	93	84	74	...	...	900	95	...
66	85	...	...	93	83	74	...	...	880	93	...
66	84	75	...	92	82	72	2.25	745	832	91	...
64	83	74	...	92	81	71	2.30	710	800	88	...
63	83	73	...	91	80	70	2.30	710	772	87	...
62	82	72	...	91	79	69	2.35	682	746	85	...
61	82	72	...	91	78	68	2.35	682	720	83	...
60	81	71	...	90	78	67	2.40	653	697	82	...
59	81	70	...	90	77	66	2.45	627	674	80	326
58	80	69	...	89	76	64	2.55	578	653	78	315
57	80	69	...	89	75	63	2.55	578	633	77	304
56	79	68	...	88	74	62	2.60	555	613	75	294
55	79	67	...	88	73	61	2.60	555	595	74	287
54	78	66	...	87	72	60	2.65	534	577	72	279
53	77	65	...	87	71	59	2.70	514	560	71	269
52	77	65	...	86	70	57	2.75	495	544	69	261
51	76	64	...	86	69	56	2.75	495	528	68	254
50	76	63	...	86	69	55	2.80	477	513	67	245
49	75	62	...	85	68	54	2.85	461	498	65	238
48	75	61	...	85	67	53	2.90	444	484	64	232
47	74	61	...	84	66	51	2.90	444	471	63	225
46	73	60	...	84	65	50	2.95	432	458	62	219
45	73	59	...	83	64	49	3.00	415	446	61	211
44	73	59	...	83	63	48	3.00	415	434	59	206
43	72	58	...	82	62	47	3.05	401	423	58	202
42	72	57	...	82	61	46	3.10	388	412	56	198
41	71	56	...	81	60	44	3.10	388	402	55	191
40	70	55	...	80	60	43	3.15	375	392	54	185
39	70	55	...	80	59	42	3.20	363	382	53	181
38	69	54	...	79	58	41	3.25	352	372	51	176
37	69	53	...	79	57	40	3.30	341	363	50	171
36	68	52	109*	78	56	39	3.35	331	354	49	168
35	68	52	109*	78	55	37	3.35	331	345	48	163
34	67	51	108*	77	54	36	3.40	321	336	46	159
33	67	50	108*	77	53	35	3.45	311	327	45	154
32	66	49	107*	76	52	34	3.50	302	318	44	150
31	66	48	106*	76	51	33	3.55	293	310	43	146
30	65	48	106*	75	50	32	3.60	285	302	42	142
29	65	47	105*	75	50	30	3.65	277	294	41	138
28	64	46	104*	74	49	29	3.70	269	286	40	134
27	64	45	103*	73	48	28	3.75	262	279	39	131
26	63	45	103*	73	47	27	3.80	255	272	38	126
25	63	44	102	72	46	26	3.80	255	266	37	124
24	62	43	101	72	45	24	3.85	248	260	37	122
23	62	42	100	71	44	23	3.90	241	254	36	118
22	62	42	99	71	43	22	3.95	235	248	35	116
21	61	41	99	70	42	21	4.00	229	243	35	113
20	61	40	98	69	42	20	4.05	223	238	34	111
18	...	...	97	...	...	...	4.10	217	230	33	107
16*	...	...	95	...	...	...	4.15	212	222	32	102
14*	...	...	94	...	...	...	4.25	203	213	31	98
12*	...	...	92	...	...	...	4.35	192	204	29	92
10*	...	...	90	...	...	...	4.40	187	195	28	90
8*	...	...	89	...	...	...	4.50	179	187	27	87
6*	...	...	87	...	...	...	4.60	170	180	26	83
4*	...	...	85	...	...	...	4.65	166	173	25	79
2*	...	...	84	...	...	...	4.80	156	166	24	77
0*	...	...	82	...	...	...	4.80	156	160	24	74
...	...	...	81	...	...	...	4.90	149	156	23	73
...	...	...	79	...	...	...	5.00	143	150	22	70
...	...	...	77	...	...	...	5.10	137	143	21	67
...	...	...	74	...	...	...	5.20	131	137	21	65
...	...	...	72	...	...	...	5.30	126	132	20	62
...	...	...	70	...	...	...	5.40	121	127	19	60
...	...	...	68	...	...	...	5.50	116	122	18	58
...	...	...	65	...	...	...	5.60	112	117	15	56

*Numbers are Rockwell "B" or "C" values not ordinarily determined.

Fig. 5-46. Hardness conversion chart permits hardness values to be converted to other scales. (Teledyne Vasco)

Hardened steel and hard alloys

C 150 Kg. "Brale" Rockwell Hardness Tester	A 60 Kg. "Brale" Rockwell Hardness Tester	D 100 Kg. "Brale" Rockwell Hardness Tester	15-N 15 Kg. N "Brale" Rockwell Superficial Hardness Tester	30-N 30 Kg. N "Brale" Rockwell Superficial Hardness Tester	45-N 45 Kg. N "Brale" Rockwell Superficial Hardness Tester	Diamond Pyramid Hardness 10 Kg. Square Base Diamond Pyramid - 136° Apex Angle	Knoop Hardness 500 Gr. & Over Knoop Elongated Diamond	Brinell Hardness 3000 Kg. Brinell Hultgren 10mm. Ball	G 150 Kg. 1/16" Ball Rockwell Hardness Tester	Tensile Strength Approx. Only Thousand lbs. per square inch
80	92.0	86.5	96.5	92.0	87.0	1865	—	—	—	INEXACT AND ONLY FOR STEEL
79	91.5	85.5	—	91.5	86.5	1787	—	—	—	
78	91.0	84.5	96.0	91.0	85.5	1710	—	—	—	
77	90.5	84.0	—	90.5	84.5	1633	—	—	—	
76	90.0	83.0	95.5	90.0	83.5	1556	—	—	—	
75	89.5	82.5	—	89.0	82.5	1478	—	—	—	
74	89.0	81.5	95.0	88.5	81.5	1400	—	—	—	
73	88.5	81.0	—	88.0	80.5	1323	—	—	—	
72	88.0	80.0	94.5	87.0	79.5	1245	—	—	—	
71	87.0	79.5	—	86.5	78.5	1160	—	—	—	
70	86.5	78.5	94.0	86.0	77.5	1076	972	—	—	
69	86.0	78.0	93.5	85.0	76.5	1004	946	—	—	
68	85.5	77.0	—	84.5	75.5	942	920	—	—	
67	85.0	76.0	93.0	83.5	74.5	894	895	—	—	
66	84.5	75.5	92.5	83.0	73.0	854	870	—	—	
65	84.0	74.5	92.0	82.0	72.0	820	846	—	—	
64	83.5	74.0	—	81.0	71.0	789	822	—	—	
63	83.0	73.0	91.5	80.0	70.0	763	799	—	—	
62	82.5	72.5	91.0	79.0	69.0	739	776	—	—	—
61	81.5	71.5	90.5	78.5	67.5	716	754	—	—	—
60	81.0	71.0	90.0	77.5	66.5	695	732	614	—	—
59	80.5	70.0	89.5	76.5	65.5	675	710	600	—	—
58	80.0	69.0	—	75.5	64.0	655	690	587	—	—
57	79.5	68.5	89.0	75.0	63.0	636	670	573	—	—
56	79.0	67.5	88.5	74.0	62.0	617	650	560	—	—
55	78.5	67.0	88.0	73.0	61.0	598	630	547	—	301
54	78.0	66.0	87.5	72.0	59.5	580	612	534	—	291
53	77.5	65.5	87.0	71.0	58.5	562	594	522	—	282
52	77.0	64.5	86.5	70.5	57.5	545	576	509	—	273
51	76.5	64.0	86.0	69.5	56.0	528	558	496	—	264
50	76.0	63.0	85.5	68.5	55.0	513	542	484	—	255
49	75.5	62.0	85.0	67.5	54.0	498	526	472	—	246
48	74.5	61.5	84.5	66.5	52.5	485	510	460	—	237
47	74.0	60.5	84.0	66.0	51.5	471	495	448	—	229
46	73.5	60.0	83.5	65.0	50.0	458	480	437	—	221
45	73.0	59.0	83.0	64.0	49.0	446	466	426	—	214
44	72.5	58.5	82.5	63.0	48.0	435	452	415	—	207
43	72.0	57.5	82.0	62.0	46.5	424	438	404	—	200
42	71.5	57.0	81.5	61.5	45.5	413	426	393	—	194
41	71.0	56.0	81.0	60.5	44.5	403	414	382	—	188
40	70.5	55.5	80.5	59.5	43.0	393	402	372	—	182
39	70.0	54.5	80.0	58.5	42.0	383	391	362	—	177
38	69.5	54.0	79.5	57.5	41.0	373	380	352	—	171
37	69.0	53.0	79.0	56.5	39.5	363	370	342	—	166
36	68.5	52.5	78.5	56.0	38.5	353	360	332	—	162
35	68.0	51.5	78.0	55.0	37.0	343	351	322	—	157
34	67.5	50.5	77.0	54.0	36.0	334	342	313	—	153
33	67.0	50.0	76.5	53.0	35.0	325	334	305	—	148
32	66.5	49.0	76.0	52.0	33.5	317	326	297	—	144
31	66.0	48.5	75.5	51.5	32.5	309	318	290	—	140
30	65.5	47.5	75.0	50.5	31.5	301	311	283	92.0	136
29	65.0	47.0	74.5	49.5	30.0	293	304	276	91.0	132
28	64.5	46.0	74.0	48.5	29.0	285	297	270	90.0	129
27	64.0	45.5	73.5	47.5	28.0	278	290	265	89.0	126
26	63.5	44.5	72.5	47.0	26.5	271	284	260	88.0	123
25	63.0	44.0	72.0	46.0	25.5	264	278	255	87.0	120
24	62.5	43.0	71.5	45.0	24.0	257	272	250	86.0	117
23	62.0	42.5	71.0	44.0	23.0	251	266	245	84.5	115
22	61.5	41.5	70.5	43.0	22.0	246	261	240	83.5	112
21	61.0	41.0	70.0	42.5	20.5	241	256	235	82.5	110
20	60.5	40.0	69.5	41.5	19.5	236	251	230	81.0	108

Soft steel, grey and malleable cast iron and most non-ferrous metal.

B 100 Kg. 1/16" Ball Rockwell Hardness Tester	F 60 Kg. 1/16" Ball Rockwell Hardness Tester	G 150 Kg. 1/16" Ball Rockwell Hardness Tester	15-T 15 Kg. 1/16" Ball Rockwell Superficial Hardness Tester	30-T 30 Kg. 1/16" Ball Rockwell Superficial Hardness Tester	45-T 45 Kg. 1/16" Ball Rockwell Superficial Hardness Tester	E 100 Kg. 1/8" Ball Rockwell Hardness Tester	K 150 Kg. 1/8" Ball Rockwell Hardness Tester	A 60 Kg. "Brale" Rockwell Hardness Tester	Knoop Hardness 500 Gr. & Over Knoop Elongated Diamond	Brinell Hardness 500 Kg. 10mm. Ball Standard Brinell	Brinell Hardness 3000 Kg. D.P.H. 10 Kg. Square Base Diamond Pyramid - 136° Apex Angle	Tensile Strength Approx. Only Thousand lbs. per square inch
100	—	82.5	93.0	82.0	72.0	—	—	61.5	251	201	240	116
99	—	81.0	92.5	81.5	71.0	—	—	61.0	246	195	234	112
98	—	79.0	—	81.0	70.0	—	—	60.0	241	189	228	109
97	—	77.5	92.0	80.5	69.0	—	—	59.5	236	184	222	106
96	—	76.0	—	80.0	68.0	—	—	59.0	231	179	216	103
95	—	74.0	91.5	79.0	67.0	—	—	58.0	226	175	210	101
94	—	72.5	—	78.5	66.0	—	—	57.5	221	171	205	98
93	—	71.0	91.0	78.0	65.5	—	—	57.0	216	167	200	96
92	—	69.0	90.5	77.5	64.5	—	100	56.5	211	163	195	93
91	—	67.5	—	77.0	63.5	—	99.5	56.0	206	160	190	91
90	—	66.0	90.0	76.0	62.5	—	98.5	55.5	201	157	185	89
89	—	64.0	89.5	75.5	61.5	—	98.0	55.0	196	154	180	87
88	—	62.5	—	75.0	60.5	—	97.0	54.0	192	151	176	85
87	—	61.0	89.0	74.5	59.5	—	96.5	53.5	188	148	172	83
86	—	59.0	88.5	74.0	58.5	—	95.5	53.0	184	145	169	81
85	—	57.5	—	73.5	58.0	—	94.5	52.5	180	142	165	80
84	—	56.0	88.0	73.0	57.0	—	94.0	52.0	176	140	162	78
83	—	54.0	87.5	72.0	56.0	—	93.0	51.0	173	137	159	77
82	—	52.5	—	71.5	55.0	—	92.0	50.5	170	135	156	75
81	—	51.0	87.0	71.0	54.0	—	91.0	50.0	167	133	153	74
80	—	49.0	86.5	70.0	53.0	—	90.5	49.5	164	130	150	72
79	—	47.5	—	69.5	52.0	—	89.5	49.0	161	128	147	Even for steel, Tensile Strength relation to hardness is inexact, unless determined for specific material.
78	—	46.0	86.0	69.0	51.0	—	88.5	48.5	158	126	144	
77	—	44.0	85.5	68.0	50.0	—	88.0	48.0	155	124	141	
76	—	42.5	—	67.5	49.0	—	87.0	47.0	152	122	139	
75	99.5	41.0	85.0	67.0	48.5	—	86.0	46.5	150	120	137	
74	99.0	39.0	—	66.0	47.5	—	85.0	46.0	147	118	135	
73	98.5	37.5	84.5	65.5	46.5	—	84.5	45.5	145	116	132	
72	98.0	36.0	84.0	65.0	45.5	—	83.5	45.0	143	114	130	
71	97.5	34.5	—	64.0	44.5	100	82.5	44.5	141	112	127	
70	97.0	32.5	83.5	63.5	43.5	99.5	81.5	44.0	139	110	125	
69	96.0	31.0	83.0	62.5	42.5	99.0	81.0	43.5	137	109	123	
68	95.5	29.5	—	62.0	41.5	98.0	80.0	43.0	135	107	121	
67	95.0	28.0	82.5	61.5	40.5	97.5	79.0	42.5	133	106	119	
66	94.5	26.5	82.0	60.5	39.5	97.0	78.0	42.0	131	104	117	
65	94.0	25.0	—	60.0	38.5	96.0	77.5	—	129	102	116	
64	93.5	23.5	81.5	59.5	37.5	95.5	76.5	41.5	127	101	114	
63	93.0	22.0	81.0	58.5	36.5	95.0	75.5	41.0	125	99	112	
62	92.0	20.5	—	58.0	35.5	94.5	74.5	40.5	124	98	110	
61	91.5	19.0	80.5	57.0	34.5	93.5	74.0	40.0	122	96	108	
60	91.0	17.5	—	56.5	33.5	93.0	73.0	39.5	120	95	107	
59	90.5	16.0	80.0	56.0	32.0	92.5	72.0	39.0	118	94	106	
58	90.0	14.5	79.5	55.0	31.0	92.0	71.0	38.5	117	92	104	
57	89.5	13.0	—	54.5	30.0	91.0	70.5	38.0	115	91	103	
56	89.0	11.5	79.0	54.0	29.0	90.5	69.5	—	114	90	101	
55	88.0	10.0	78.5	53.0	28.0	90.0	68.5	37.5	112	89	100	
54	87.5	8.5	—	52.5	27.0	89.5	68.0	37.0	111	87	—	
53	87.0	7.0	78.0	51.5	26.0	89.0	67.0	36.5	110	86	—	
52	86.5	5.5	77.5	51.0	25.0	88.0	66.0	36.0	109	85	—	
51	86.0	4.0	—	50.5	24.0	87.5	65.0	35.5	108	84	—	
50	85.5	2.5	77.0	49.5	23.0	87.0	64.5	35.0	107	83	—	

B 100 Kg. 1/16" Ball Rockwell Hardness Tester	F 60 Kg. 1/16" Ball Rockwell Hardness Tester	15-T 15 Kg. 1/16" Ball Rockwell Superficial Hardness Tester	30-T 30 Kg. 1/16" Ball Rockwell Superficial Hardness Tester	45-T 45 Kg. 1/16" Ball Rockwell Superficial Hardness Tester	E 100 Kg. 1/8" Ball Rockwell Hardness Tester	H 60 Kg. 1/8" Ball Rockwell Hardness Tester	K 150 Kg. 1/8" Ball Rockwell Hardness Tester	A 60 Kg. "Brale" Rockwell Hardness Tester	Knoop Hardness 500 Gr. & Over Knoop Elongated Diamond
50	85.5	77.0	49.5	23.0	87.0	—	64.5	35.0	107
49	85.0	76.5	49.0	22.0	86.5	—	63.5	—	106
48	84.5	—	48.5	20.5	85.5	—	62.5	34.5	105
47	84.0	76.0	47.5	19.5	85.0	—	61.5	34.0	104
46	83.0	75.5	47.0	18.5	84.5	—	61.0	33.5	103
45	82.5	—	46.0	17.5	84.0	—	60.0	33.0	102
44	82.0	75.0	45.5	16.5	83.5	—	59.0	32.5	101
43	81.5	74.5	45.0	15.5	82.5	—	58.0	32.0	100
42	81.0	—	44.0	14.5	82.0	—	57.5	31.5	99
41	80.5	74.0	43.5	13.5	81.5	—	56.5	31.0	98
40	79.5	73.5	43.0	12.5	81.0	—	55.5	—	97
39	79.0	—	42.0	11.0	80.0	—	54.5	30.5	96
38	78.5	73.0	41.5	10.0	79.5	—	54.0	30.0	95
37	78.0	72.5	40.5	9.0	79.0	—	53.0	29.5	94
36	77.5	—	40.0	8.0	78.5	100	52.0	29.0	93
35	77.0	72.0	39.5	7.0	78.0	99.5	51.5	28.5	92
34	76.5	71.5	38.5	6.0	77.0	99.0	50.5	28.0	91
33	75.5	—	38.0	5.0	76.5	—	49.5	—	90
32	75.0	71.0	37.5	4.0	76.0	98.5	48.5	27.5	89
31	74.5	—	36.5	3.0	75.5	98.0	48.0	27.0	88
30	74.0	70.5	36.0	2.0	75.0	—	47.0	26.5	—
29	73.5	70.0	35.5	1.0	74.0	97.5	46.0	26.0	—
28	73.0	—	34.5	—	73.5	97.0	45.0	25.5	—
27	72.5	69.5	34.0	—	73.0	96.5	44.5	25.0	85
26	72.0	69.0	33.0	—	72.5	—	43.5	24.5	—
25	71.0	—	32.5	—	72.0	96.0	42.5	—	—
24	70.5	68.5	32.0	—	71.0	95.5	41.5	24.0	—
23	70.0	68.0	31.0	—	70.5	—	41.0	23.5	82
22	69.5	—	30.5	—	70.0	95.0	40.0	23.0	—
21	69.0	67.5	29.5	—	69.5	94.5	39.0	22.5	—
20	68.5	—	29.0	—	68.5	—	38.0	22.0	—
19	68.0	67.0	28.5	—	68.0	94.0	37.5	21.5	79
18	67.0	66.5	27.5	—	67.5	93.5	36.5	—	—
17	66.5	—	27.0	—	67.0	93.0	35.5	21.0	—
16	66.0	66.0	26.0	—	66.5	—	35.0	20.5	—
15	65.5	65.5	25.5	—	65.5	92.5	34.0	20.0	76
14	65.0	—	25.0	—	65.0	92.0	33.0	—	—
13	64.5	65.0	24.0	—	64.5	—	32.0	—	—
12	64.0	64.5	23.5	—	64.0	91.5	31.5	—	—
11	63.5	—	23.0	—	63.5	91.0	30.5	—	73
10	63.0	64.0	22.0	—	62.5	90.5	29.5	—	—
9	62.0	—	21.5	—	62.0	—	29.0	—	—
8	61.5	63.5	20.5	—	61.5	90.0	28.0	—	71
7	61.0	63.0	20.0	—	61.0	89.5	27.0	—	—
6	60.5	—	19.5	—	60.5	—	26.0	—	—
5	60.0	62.5	18.5	—	60.0	89.0	25.5	—	69
4	59.5	62.0	18.0	—	59.0	88.5	24.5	—	—
3	59.0	—	17.0	—	58.5	88.0	23.5	—	—
2	58.0	61.5	16.5	—	58.0	—	23.0	—	68
1	57.5	61.0	16.0	—	57.5	87.5	22.0	—	—
0	57.0	—	15.0	—	57.0	87.0	21.0	—	67

Although conversion tables dealing with hardness can only be approximate and never mathematically exact, it is of considerable value to be able to compare different hardness scales.

This table is based on the assumption that the metal tested is homogeneous to a depth several times as great as the depth of the indentation.

In metal not homogeneous, different loads and different shapes of penetrators would penetrate, or at least meet the resistance of, metal of varying hardness, depending upon the depth of the indentation. Therefore, no recorded hardness value could be confirmed by another person unless shape of penetrator and actual load applied are both specified.

The indentation hardness values measured on the various scales depend on the work hardening behavior of the material during the test and this in turn depends on the degree of previous cold working of the material. The B-scale relationships in the table are based largely on annealed metals for the low values and cold worked metals for the higher values. Therefore, annealed metals of high B-scale hardness such as austenitic stainless steels, nickel and high nickel alloys do not conform closely to these general tables. Neither do cold-worked metals of low B-scale hardness such as aluminum and the softer alloys. Special correlations are needed for more exact relationships in these cases.

Fig. 5-47. Hardness conversion chart compares different hardness scales. (Acco Industries, Wilson Instrument Div.)

ROCKWELL				SUPERFICIAL ROCKWELL		
Diamond Brale			1 16" Ball	"N" Brale Penetrator		
150 kgm C Scale	60 kgm A Scale	100 kgm D Scale	100 kgm B Scale	15 kg Load 15 N	30 kg Load 30 N	45 kg Load 45 N
68	86	...	...	93	84	75
68	85	...	...	93	84	75
67	85	...	...	93	84	74
66	85	...	...	93	83	74
66	84	75	...	92	82	72
64	83	74	...	92	81	71
63	83	73	...	91	80	70
62	82	72	...	91	79	69
61	82	72	...	91	78	68
60	81	71	...	90	78	67
59	81	70	...	90	77	66
58	80	69	...	89	76	64
57	80	69	...	89	75	63
56	79	68	...	88	74	62
55	79	67	...	88	73	61
54	78	66	...	87	72	60
53	77	65	...	87	71	59
52	77	65	...	86	70	57
51	76	64	...	86	69	56
50	76	63	...	86	69	55
49	75	62	...	85	68	54
48	75	61	...	85	67	53
47	74	61	...	84	66	51
46	73	60	...	84	65	50
45	73	59	...	83	64	49
44	73	59	...	83	63	48
43	72	58	...	82	62	47
42	72	57	...	82	61	46
41	71	56	...	81	60	44
40	70	55	...	80	60	43
39	70	55	...	80	59	42
38	69	54	...	79	58	41
37	69	53	...	79	57	40
36	68	52	109*	78	56	39
35	68	52	109*	78	55	37
34	67	51	108*	77	54	36
33	67	50	108*	77	53	35

Fig. 5-48. Chart shows conversion from Rockwell "C" scale to Rockwell "A" scale. (Teledyne Vasco)

scale is found to be the equivalent of 77 R_A on the "A" scale in Fig. 5-48. If a blueprint calls for a hardness of 235 BHN (Brinnel) and you have only a Rockwell testing machine, the material should test out at approximately 99 R_B on the "B" scale or 22 R_C on the "C" scale. See Fig. 5-49.

Also, a hardness of 321 BHN (Brinnel) is the same hardness as 46 *Shore* (Scleroscope) or 36-45N on the Rockwell Superficial tester. See Fig. 5-50.

The larger the numerical value of hardness, the harder the part is. This is true of every hardness scale on the charts. Therefore, as we move toward the top of the charts, the hardness values are harder. If we move toward the bottom of the charts, the hardness of a material is softer. For example, a hardness of 28 R_C would be slightly higher than 35 units *Shore*. A hardness of 432 BHN would be slightly higher than 434 DPH (Vickers). Consider these values. Which would be harder, 30 R_C or 302 BHN or 49 *Shore*? Many interesting problems can be made up to give you practice at reading these charts.

ROCKWELL				SUPERFICIAL ROCKWELL			BRINELL		Vickers of Firth Diamond Hardness Number
Diamond Brale			1 16" Ball	"N" Brale Penetrator			10 m/m Ball 3000 kgm Load		
150 kgm C Scale	60 kgm A Scale	100 kgm D Scale	100 kgm B Scale	15 kg Load 15 N	30 kg Load 30 N	45 kg Load 45 N	Diam. of Ball Im- pression in m/m	Hardness Number	
68	86	...	...	93	84	75	...	...	940
68	85	...	...	93	84	75	...	...	920
67	85	...	...	93	84	74	...	...	900
66	85	...	...	93	83	74	...	...	880
66	84	75	...	92	82	72	2.25	745	832
64	83	74	...	92	81	71	2.30	710	800
63	83	73	...	91	80	70	2.30	710	772
62	82	72	...	91	79	69	2.35	682	746
61	82	72	...	91	78	68	2.35	682	720
60	81	71	...	90	78	67	2.40	653	697
59	81	70	...	90	77	66	2.45	627	674
58	80	69	...	89	76	64	2.55	578	653
57	80	69	...	89	75	63	2.55	578	633
56	79	68	...	88	74	62	2.60	555	613
55	79	67	...	88	73	61	2.60	555	595
54	78	66	...	87	72	60	2.65	534	577
53	77	65	...	87	71	59	2.70	514	560
52	77	65	...	86	70	57	2.75	495	544
51	76	64	...	86	69	56	2.75	495	528
50	76	63	...	86	69	55	2.80	477	513
49	75	62	...	85	68	54	2.85	461	498
48	75	61	...	85	67	53	2.90	444	484
47	74	61	...	84	66	51	2.90	444	471
46	73	60	...	84	65	50	2.95	432	458
45	73	59	...	83	64	49	3.00	415	446
44	73	59	...	83	63	48	3.00	415	434
43	72	58	...	82	62	47	3.05	401	423
42	72	57	...	82	61	46	3.10	388	412
41	71	56	...	81	60	44	3.10	388	402
40	70	55	...	80	60	43	3.15	375	392
39	70	55	...	80	59	42	3.20	363	382
38	69	54	...	79	58	41	3.25	352	372
37	69	53	...	79	57	40	3.30	341	363
36	68	52	109*	78	56	39	3.35	331	354
35	68	52	109*	78	55	37	3.35	331	345
34	67	51	108*	77	54	36	3.40	321	336
33	67	50	108*	77	53	35	3.45	311	327
32	66	49	107*	76	52	34	3.50	302	318
31	66	48	106*	76	51	33	3.55	293	310
30	65	48	106*	75	50	32	3.60	285	302
29	65	47	105*	75	50	30	3.65	277	294
28	64	46	104*	74	49	29	3.70	269	286
27	64	45	103*	73	48	28	3.75	262	279
26	63	45	103*	73	47	27	3.80	255	272
25	63	44	102	72	46	26	3.80	255	266
24	62	43	101	72	45	24	3.85	248	260
23	62	42	100	71	44	23	3.90	241	254
22	62	42	99	71	43	22	3.95	235	248
21	61	41	99	70	42	21	4.00	229	243
20	61	40	98	69	42	20	4.05	223	238
18	...	...	97	...	...	...	4.10	217	230
16*	...	...	95	...	...	...	4.15	212	222
14*	...	...	94	...	...	...	4.25	203	213
12*	...	...	92	...	...	...	4.35	192	204

Fig. 5-49. Chart indicates conversion from BHN scale to Rockwell "B" and Rockwell "C" scale. (Teledyne Vasco)

SUPERFICIAL ROCKWELL			BRINELL 10 m/m Ball 3000 kgm Load		Vickers of Firth Diamond Hardness Number	Sclero-scope	Tensile Strength
"N" Brale Penetrator							
15 kg Load 15 N	30 kg Load 30 N	45 kg Load 45 N	Diam. of Ball Im-pression in m/m	Hardness Number			Equivalent 1000 Lb. Sq. In.
93	84	75	...	...	940	97	...
93	84	75	...	...	920	96	...
93	84	74	...	...	900	95	...
93	83	74	...	...	880	93	...
92	82	72	2.25	745	832	91	...
92	81	71	2.30	710	800	88	...
91	80	70	2.30	710	772	87	...
91	79	69	2.35	682	746	85	...
91	78	68	2.35	682	720	83	...
90	78	67	2.40	653	697	82	...
90	77	66	2.45	627	674	80	326
89	76	64	2.55	578	653	78	315
89	75	63	2.55	578	633	77	304
88	74	62	2.60	555	613	75	294
88	73	61	2.60	555	595	74	287
87	72	60	2.65	534	577	72	279
87	71	59	2.70	514	560	71	269
86	70	57	2.75	495	544	69	261
86	69	56	2.75	495	528	68	254
86	69	55	2.80	477	513	67	245
85	68	54	2.85	461	498	65	238
85	67	53	2.90	444	484	64	232
84	66	51	2.90	444	471	63	225
84	65	50	2.95	432	458	62	219
83	64	49	3.00	415	446	61	211
83	63	48	3.00	415	434	59	206
82	62	47	3.05	401	423	58	202
82	61	46	3.10	388	412	56	198
81	60	44	3.10	388	402	55	191
80	60	43	3.15	375	392	54	185
80	59	42	3.20	363	382	53	181
79	58	41	3.25	352	372	51	176
79	57	40	3.30	341	363	50	171
78	56	39	3.35	331	354	49	168
78	55	37	3.35	331	345	48	163
77	54	36	3.40	321	336	46	159
77	53	35	3.45	311	327	45	154
76	52	34	3.50	302	318	44	150
76	51	33	3.55	293	310	43	146
75	50	32	3.60	285	302	42	142
75	50	30	3.65	277	294	41	138
74	49	29	3.70	269	286	40	134
73	48	28	3.75	262	279	39	131
73	47	27	3.80	255	272	38	126
72	46	26	3.80	255	266	37	124

Fig. 5-50. Chart serves to convert BHN scale to Shore and 45N Rockwell Superficial. (Teledyne Vasco)

TEST YOUR KNOWLEDGE

Write your answers on a separate sheet of paper. Do not write in this book.

Use the following words to answer the 19 questions below. Some of these words may be used more than once, some of the words may not be used at all.

Brinell   Moh   Sonodur
File   Rockwell   Superficial
Knoop   Shore   Vickers

1. In which hardness testing method is the hardness based on the diameter of an indenture?
2. In which hardness testing method does the impression look like a diamond, wherein one axis of the diamond is seven times as long as the other?
3. What hardness testing method is the most widely used of all methods?
4. In which hardness testing method is the hardness value dependent on the height of the bounce?
5. In which two hardness testing methods is the hardness dependent on the depth of penetration rather than the width of penetration?
6. In which hardness testing method is a diamond-tipped magnetostrictive rod used?
7. In which hardness testing methods is a minor load applied first to get through the outer crust or the outer surface of the metal before making the hardness test?
8. In most hardness testing methods, the surface should be as horizontal and as flat and smooth as possible. In which hardness testing method is this the most critical?
9. In which type of hardness testing method is the hardness dependent on the width of a square based diamond shaped impression?
10. In which hardness testing method is a very small impression made by a diamond penetrator, due to a load that is so small it is measured in grams instead of kilograms? A small microscope is also used in this method.
11. What hardness testing method employs either a C, A, E, G, or K scale?
12. More than one hardness testing method measures hardness in units of BHN. List the oldest method that does this.
13. "Microhardness tester" is another name for which hardness testing method?
14. What hardness testing method utilizes a 10 millimeter diameter ball and a 3 000 kilogram force?
15. The 30T scale refers to which hardness testing method?
16. Name the hardness testing method that measures in units of DPH.
17. Name the hardness testing method that involves scratching a surface with ten stones.
18. Name a hardness testing method that employs either a diamond point penetrator or a ball, and it employs a load of 15, 30, or 45 kilograms.
19. What hardness testing method is the fastest to use, but does not give accurate numerical results?

The following questions also involve hardness testing:

20. In the 45T scale, what does the 45 stand for?
21. What is the greatest hardness value, in Rockwell units, that steel can attain approximately?
22. What Rockwell hardness scale is most commonly used for hard materials?
23. In the Rockwell hardness test, what scales use the diamond point penetrator?

Use the hardness conversion charts in Figs. 5-46 and 5-47 to determine the answers to the following questions.

24. Which of the following values would represent the hardest material?
   a. 35 Rockwell C.
   b. 69 Rockwell A.
   c. 44 Rockwell D.
25. Which of the following values would represent the hardest material?

a. 460 BHN.

b. 53 Rockwell C.

c. 271 DPH (Vickers).

26. Which of the following values would represent the hardest material?

a. 38 *Shore*.

b. 25 Rockwell C.

c. 97 Rockwell B.

27. Which of the following values would represent the hardest material?

a. 389 DPH (Vickers).

b. 61 *Shore*.

c. 389 BHN.

28. Which of the following values would represent the hardest material?

a. 426 *Knoop*.

b. 30 Rockwell C.

c. 270 BHN.

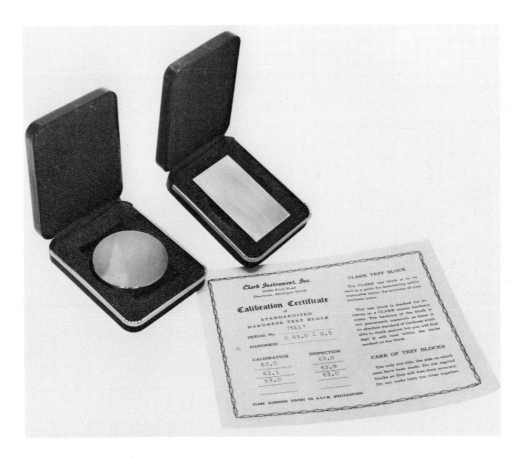

Hardness testing machines are periodically calibrated, using test blocks similar to those pictured above. Periodic calibration is necessary so that hardness testing machines will read to precise numerical hardness values. (Clark Instrument, Inc.)

6 PROPERTIES OF STEEL

After studying this chapter, you will be able to:

☐ Tell what is meant by the term "a property of steel."

☐ Name and describe three key mechanical properties.

☐ Explain what ductility is and why it is important to steel.

☐ Discuss the chemical, electrical, and thermal properties of metal.

☐ Use a chart to compare the properties of metals with each other.

PROPERTIES OF METALS

You may have heard the expression "properties of a metal." Properties of metals refer to their characteristics, abilities, special traits, strengths, advantages, disadvantages, unusual features, and to how they compare to other metals. Just as a human being may have the strength to lift heavy boxes or the ability to resist smoking cigarettes, similarly, a metal may exhibit strength to carry large loads or the ability to resist decay from corrosion.

Every material can be described by:
1. Its MECHANICAL PROPERTIES, such as strength and hardness.
2. Its CHEMICAL PROPERTIES, such as corrosion resistance.
3. Its ELECTRICAL PROPERTIES, such as resistivity.
4. Its THERMAL PROPERTIES, such as melting temperature.
5. OTHER PROPERTIES, such as weight and cost. See Fig. 6-1.

THREE KEY MECHANICAL PROPERTIES

The three properties that are discussed most often in metallurgy are HARDNESS, STRENGTH, and BRITTLENESS. They are like three inseparable brothers. They change together.

IMPORTANT PROPERTIES IN STEEL				
Mechanical Properties	Chemical Properties	Electrical Properties	Thermal Properties	Other Properties

Hardness
Tensile Strength
Compressive Strength
Shear Strength
Fatigue Strength
Toughness
Impact Strength
Shock Resistance
Flexure Strength
Brittleness
Ductility
Percent Elongation
Wear
	Corrosion
	  Resistance
	Resistance to
	  Acids
	Resistance to
	  Alkali
	Resistance to
	  Other Chemicals
		Electrical
		  Conductivity
		Electrical
		  Resistance
		Dielectric
		  Strength
			Coefficient of
			  Thermal
			  Expansion
			Melting
			  Temperature
			Thermal
			  Conductivity
				Cost
				Weight
				Availability

Fig. 6-1. Steel has many important properties.

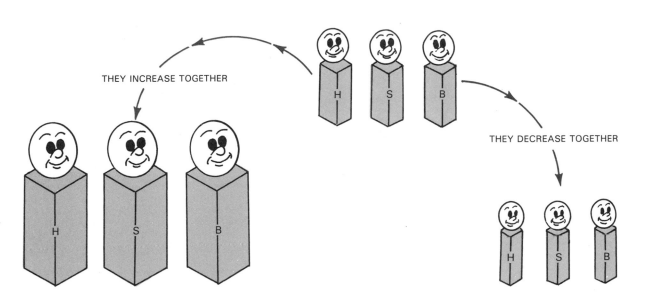

Fig. 6-2. Hardness (H), strength (S), and brittleness (B) change together.

See Fig. 6-2. As one increases in value, generally, the other two also increase. As one decreases, generally, the other two also decrease.

Strength and hardness are normally desirable properties in metal. Brittleness is generally a bad characteristic. Thus, the problem in metallurgy becomes: how can you increase the hardness and strength of a material without also increasing its brittleness?

HARDNESS

Only when special alloys are added to metal, can the hardness and the strength of a metal be improved without increasing the brittleness. Millions of dollars of metallurgical research have gone into the development of steels that can increase their hardness and strength while not increasing their brittleness.

Hardness was discussed in Chapter 5. Strength and brittleness are discussed in this chapter.

STRENGTH

There are many different types of STRENGTH. They are: tensile strength, compressive strength, shear strength, fatigue strength, impact strength, and flexure strength.

TENSILE STRENGTH is a metal's ability to withstand stress in tension. Tension is a "pulling apart," Fig. 6-3. This is perhaps the most important of all the strengths. Steel is very strong in tension.

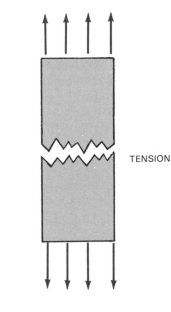

TENSION

Fig. 6-3. Tension is a pulling apart.

COMPRESSIVE STRENGTH is the ability to withstand a pressing or "squeezing together," Fig. 6-4. Cast iron has outstanding compressive strength.

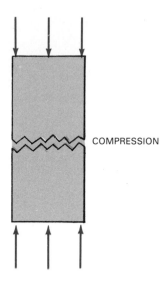

Fig. 6-4. Compression is a squeezing together.

MATERIAL	TENSILE STRENGTH(psi)	COMPRESSIVE STRENGTH (psi)
1025 Steel	70,000	70,000
1095 Steel	110,000	110,000
52100 Steel	140,000	140,000
Gray Cast Iron	35,000	110,000
Wrought Iron	40,000	40,000
Stainless Steel	95,000	95,000
Aluminum	40,000	40,000
Bronze	60,000	60,000
Zinc	20,000	20,000

Fig. 6-5. Cast iron has greater compression strength than tensile strength.

Most materials have approximately equal abilities in resisting tension and in resisting compression. A few materials, such as cast iron and concrete, are able to take much higher compressive stresses than tensile stresses, Fig. 6-5.

SHEAR STRENGTH is the ability to resist a "sliding past" type of action, Fig. 6-6. The ability to take shear stress is less in most materials. However, the shear stresses that are applied to materials are, by nature, generally also less.

FATIGUE STRENGTH or ENDURANCE STRENGTH refers to the ability to take a repeated loading. A material will fail at a lower strength level if a force is continually applied and withdrawn, applied and withdrawn, applied and withdrawn. Vibration produces fatigue stress. Some materials will hold up to a steady tensile stress or compressive stress without breaking, but will fail quickly under a small fatigue stress. Such a material has little ability to withstand repeated loading and unloading. Slight cracks in the surface tend to grow and work their

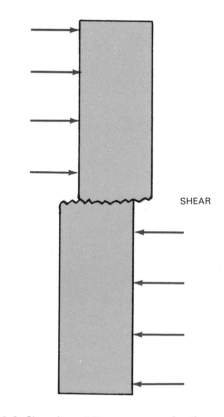

Fig. 6-6. Shear is a sliding past type of action.

way across a piece of metal when it lacks good fatigue strength, Fig. 6-7. If you repeatedly fold and crease a piece of paper before trying to rip it, it will "give up" much more easily.

TOUGHNESS and IMPACT STRENGTH measure the ability to resist shock. A material must have a good combination of both strength and ductility to resist shock. Air hammers, connecting rods in engines, and impact wrenches all must resist shock. Therefore, an important property in them is toughness.

Some materials can resist high forces or loads if the loads are applied gradually and gently. But some of these same materials cannot tolerate even a small force if it is suddenly applied. The science of karate illustrates that. Strong materials can be broken with susprising ease if the speed of the blow is quite rapid.

To cite some examples, a material like cast iron, which has good strength but poor ductility, does not have good shock resistance. Medium carbon steel has fairly good strength and ductility. It has good toughness and shock resistance.

FLEXURE STRENGTH is bending strength. It generally involves tension on one side of a material and compression on the opposite side, Fig. 6-8. This is regularly encountered in beams and long parts in machines.

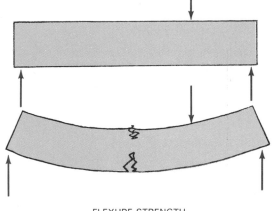

FLEXURE STRENGTH

Fig. 6-8. Flexure strength involves tension on one side and compression on the opposite side.

BRITTLENESS AND DUCTILITY

BRITTLENESS and DUCTILITY are opposites. If a material stretches very much before it breaks, it is said to be DUCTILE or to have high ductility. If a material does not stretch before it fractures, it is said to be BRITTLE or have a high degree of brittleness, Fig. 6-9.

Almost always, ductility is more desirable than brittleness, because a ductile material is able to resist shock better. However, if a slight deforming of the material causes dimensional trouble, ductility is no longer an asset.

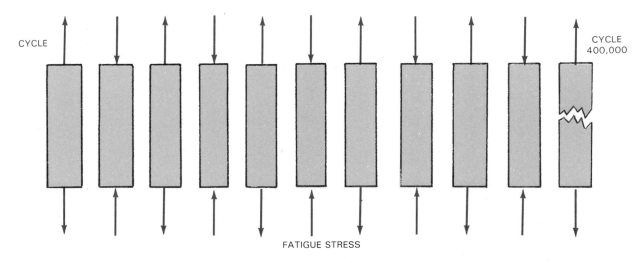

CYCLE

CYCLE 400,000

FATIGUE STRESS

Fig. 6-7. Fatigue strength deals with repeated loading and unloading.

113

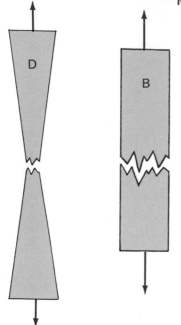

Fig. 6-9. A ductile (D) material stretches much more before it fractures than a brittle (B) material does.

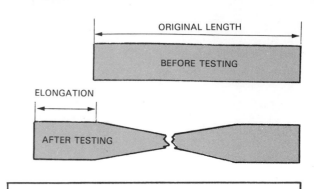

% ELONGATION = $\dfrac{ELONGATION}{ORIGINAL\ LENGTH} \times 100$

Fig. 6-10. Percent elongation is the percent the material will stretch before it breaks.

All brittle materials are not always stronger than all ductile materials. A ductile material may resist high forces while stretching.

Low carbon steel, aluminum, and rubber bands are ductile. Cast iron, glass, and uncooked spaghetti are brittle.

A common measure of the ductility of a material is called PERCENT ELONGATION. This means the percent that a material will stretch before it breaks, Fig. 6-10.

WEAR

WEAR is a very important property of a material. In many applications, a few thousandths of an inch of wear can cause an entire machine to fail.

Wear is the ability of a material to withstand wearing away. It is the ability of a metal to resist a slow deterioration, usually over a long period of time, that is caused by a frictional scratching, scoring, galling, scuffing, or seizing, Fig. 6-11. Wear is also caused by pitting, Fig. 6-12, or by fretting, Fig. 6-13.

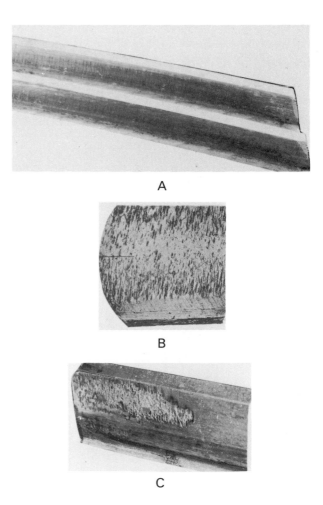

Fig. 6-11. Many types of wear can occur in a metal: A — Normal wear (surface polishing). B — Abrasive wear (scratching). C — Adhesive wear (scoring, galling, scuffing, seizing). (The Falk Corporation, subsidiary of Sundstrand Corporation)

114

Proper lubrication is effective in reducing wear. Figs. 6-14, and 6-15 show the effects of improper and proper lubrication on a typical engine bearing insert.

Fig. 6-12. Another type of wear that can occur in a metal is pitting. (The Falk Corporation, subsidiary of Sundstrand Corporation)

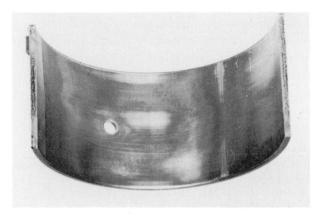

Fig. 6-14. A typical used engine bearing insert shows normal wear. (Texaco Inc.)

Fig. 6-13. Still another type of wear that can occur in a metal is fretting.
(The Falk Corporation, subsidiary of Sundstrand Corporation)

The ability to resist wear is highly dependent on the hardness of the material. The harder a material becomes generally determines how great its ability will be to resist wear.

A

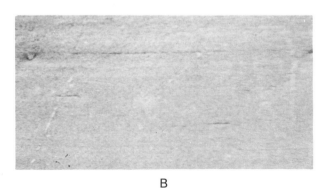

B

Fig. 6-15. Close-ups of used engine bearing inserts reveal: A — Surface damaged by ineffective lubrication. B — Surface that shows very little wear because it was properly lubricated. These photographs are magnified 25X. (Texaco Inc.)

CHEMICAL PROPERTIES

CORROSION RESISTANCE is perhaps the most important chemical property of a metal. A metal that has good corrosion resistance is able to protect itself against chemical attack by the environment. A corrosion resistant material will resist humidity without deteriorating. It will also resist sunlight, water, and heat.

Steel or iron in their natural state tend to rust when they come into contact with a humid atmosphere or with water. Rust is iron oxide, FeO. By adding special alloys to the steel (which increases the price of the steel), good corrosion resistance can be attained.

There are many types of corrosion. Some examples are shown in Figs. 6-16 through 6-19.

Other important chemical properties include RESISTANCE TO ACIDS, RESISTANCE TO ALKALI and SALTS, and RESISTANCE TO OTHER CHEMICALS.

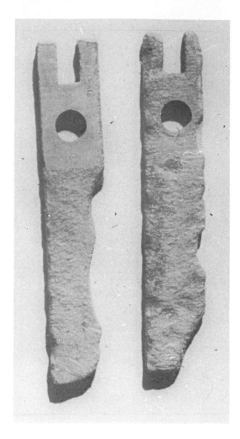

Fig. 6-17. Close-up shows erosion corrosion of cast iron parts. (The International Nickel Company, Inc.)

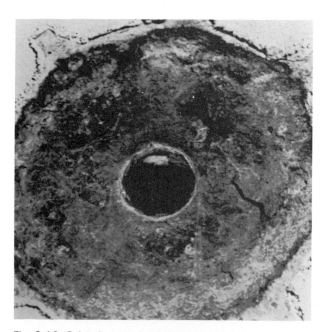

Fig. 6-16. Galvanic corrosion has occured where magnesium is in close contact with steel core. (The International Nickel Company, Inc.)

Fig. 6-18. Small pits distributed at random are evidence of pitting corrosion. (The International Nickel Company, Inc.)

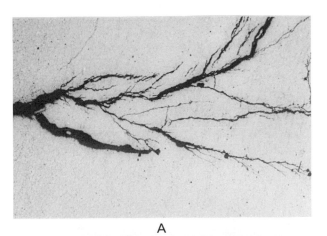

A

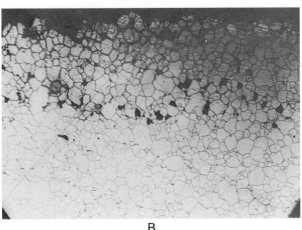

B

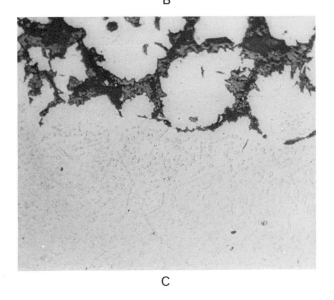

C

Fig. 6-19. Microscopic examination reveals: A — Transcrystalline stress-corrosion of stainless steel under 100x magnification. B — Intergranular stress-corrosion of steel under 100x magnification. C — Carburization and oxidation of steel at high temperatures under 250x magnification. (The International Nickel Company, Inc.)

ELECTRICAL PROPERTIES

If electricity can flow freely through a material, we say that the material has high ELECTRICAL CONDUCTIVITY. If the material refuses to let electricity flow through it, we say it has high ELECTRICAL RESISTANCE. Steel has a very high electrical conductivity and a low resistance to electrical flow, Fig. 6-20.

DIELECTRIC STRENGTH is another popular electrical property. A material with good dielectric strength is able to withstand a large voltage over a prolonged time period without passing current or breaking down, Fig. 6-21.

THERMAL PROPERTIES

Whenever any metal is heated, it grows larger. Some metals expand more rapidly than others as the temperature increases, Fig. 6-22. The COEFFICIENT OF THERMAL EXPANSION is a term which describes how fast a material expands when it is subjected to heat. In Fig. 6-23, the coefficient of thermal expansion is compared for different materials.

MELTING POINT is another important thermal property of metal. This is the temperature at which the material will change from a solid to a liquid. For steel, this temperature is in the vicinity of 3000°F. The temperatures at which ten different materials melt are listed on the chart in Fig. 6-24.

If heat can travel rapidly across a material, it is said to have a high degree of THERMAL CONDUCTIVITY. In this regard, then, a material that has a high degree of thermal conductivity will transmit heat rapidly. If you are trying to get rid of heat, this property is an advantage. If you are trying to retain heat, this trait is a disadvantage.

Aluminum and copper both have high degrees of thermal conductivity. Steel is about average for a metal. Gray iron, steel, and other materials are compared in Fig. 6-25.

STEEL AND OTHER MATERIALS WITH HIGH
ELECTRIC CONDUCTIVITY

MATERIALS WITH HIGH ELECTRIC RESISTANCE

Fig. 6-20. Electrical conductivity and electrical resistance are two important properties of steel.

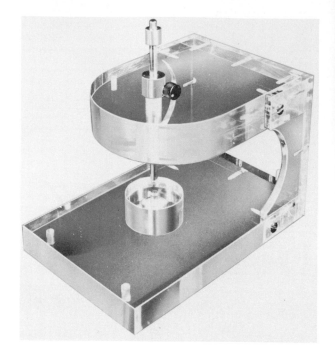

Fig. 6-21. This electrode assembly is used to test dielectric strength. (Beckman Instruments, Inc., Cedar Grove Operations)

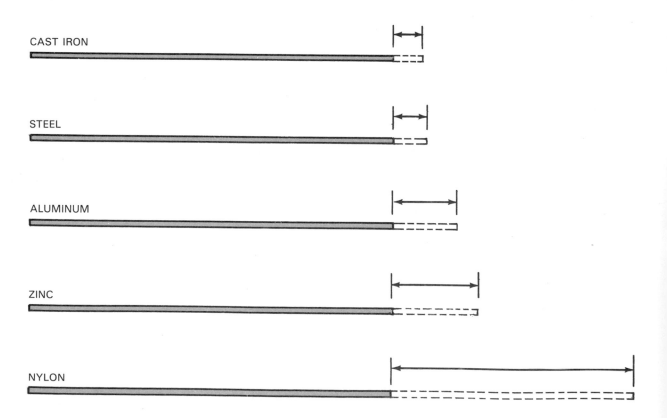

CAST IRON

STEEL

ALUMINUM

ZINC

NYLON

Fig. 6-22. Different materials increase in length more rapidly than others as the temperature increases. The arrows indicate comparative increases in thermal expansion between five different materials.

MATERIAL	COEFFICIENT OF THERMAL EXPANSION PER DEGREES F
Gray Iron	6.0×10^{-6}
Steel	6.5×10^{-6}
Nickel	7.3×10^{-6}
Copper	9.2×10^{-6}
Bronze	10.0×10^{-6}
Brass	10.3×10^{-6}
Aluminum	12.8×10^{-6}
Magnesium	14.4×10^{-6}
Zinc	17.0×10^{-6}
Nylon	50.0×10^{-6}
Polystyrene	100.0×10^{-6}

Fig. 6-23. Table compares "coefficient of thermal expansion" for different materials.

MATERIAL	THERMAL CONDUCTIVITY ($CAL/CM^2/^\circ C/SEC/CM$)
Gray Iron	.11
Steel	.11
Nickel	.22
Copper	.94
Aluminum	.45
Magnesium	.37
Zinc	.27
Lead	.08
Nylon	.0006
Polystyrene	.001

Fig. 6-25. Table compares "thermal conductivity" for different materials.

MATERIAL	MELTING TEMPERATURE ($^\circ F$)
Gray Iron	2400
Steel	2700
Nickel	2650
Copper	1980
Aluminum	1220
Magnesium	1200
Zinc	790
Lead	620
Nylon	300
Polystyrene	250

Fig. 6-24. Table compares "melting temperature" for different materials.

OTHER PROPERTIES

The COST of steel is relatively low compared to most other popular metals. This low cost factor combined with its good strength is the reason that steel is the most widely used metal in the world.

WEIGHT is another property that often is of importance. Here aluminum or magnesium have strong advantages over steel if light weight is an important factor, Fig. 6-26.

AVAILABILITY of steel is not a property of the metal itself, but it is a factor that has to be taken into consideration when selecting a material to do a job. Steel is widely available and is sold in all major cities. It usually is more readily available than any other metal.

MATERIAL	WEIGHT (#/FT³)
Gray Iron	482
Steel	490
Nickel	550
Copper	555
Aluminum	170
Magnesium	109
Zinc	440
Lead	710
Nylon	70
Polystyrene	60

Fig. 6-26. Table compares "weight" of different materials.

COMPARISON CHARTS
OF METAL PROPERTIES

It is interesting to compare the properties of different metals. Fig. 6-27 compares most of these properties for steel, cast iron, wrought iron, aluminum, copper, bronze, brass, zinc, lead, nickel, tin, titanium, and tungsten.

Some materials are superior to steel in one particular property. However, the overall combination of good properties is what makes steel one of the most widely used materials in the world today.

MATERIAL	TENSILE STRENGTH "psi x 10⁻³"	COMPRESSIVE STRENGTH "psi x 10⁻³"	HARDNESS "BHN"	MODULUS OF ELASTICITY "psi x 10⁻⁶"	WEIGHT "LBS/FT³"
Steel	60-200	Same as Tensile	150-620	30	490
Cast Iron	20-100	80-180	140-325	15	482
Wrought Iron	40	Same as Tensile	NA	28	490
Aluminum	20-60	''	50-110	10	170
Copper	30-60	''	40	17	555
Bronze	65-130	''	100-200	17	550
Brass	30-100	''	50-160	15	520
Zinc	20-30	NA	80-90	NA	440
Lead	2-5	NA	5-12	20	710
Nickel	45-60	NA	80-380	30	550
Tin	3-9	NA	7	6	450
Titanium	50-135	NA	NA	16	280
Tungsten	220	NA	NA	59	1180

MATERIAL	ELECTRIC CONDUCTIVITY "100 SILVER COMPARISON SCALE"	ELECTRIC RESISTANCE "MICRO-OHM/CM"	COEFFICENT OF THERMAL EXPANSION "IN/IN°Fx10⁶"	MELTING TEMPERATURE "°F"	THERMAL CONDUCTIVITY "CAL/CM²/°C/ SEC/CM"
Steel	12	17	6.3	2700	.11
Cast Iron	5	30	6.5	2400	.11
Wrought Iron	15	NA	6.6	2800	.18
Aluminum	63	3	12	1220	.45
Copper	98	2	9	1980	.94
Bronze	36	7	10	1840	.29
Brass	28	6.5	10	1700	.28
Zinc	30	6	17	790	.27
Lead	8.4	23	16	620	.08
Nickel	13	20	7	2650	.22
Tin	14	11.5	15	450	.16
Titanium	14	120	5	3300	.03
Tungsten	14	5.5	2.4	6100	.48

Fig. 6-27. Table compares many properties of different metals. Values listed may vary for different alloys and compositions of each material. Cost is not included because it fluctuates rapidly for some of the materials listed.

TEST YOUR KNOWLEDGE

Write your answers on a separate sheet of paper. Do not write in this book.

1. Three important properties of a metal, that generally change together are _____, _____, and _____.
2. Which of these three properties are generally considered desirable?
3. Which of these three properties are generally considered undesirable?
4. What type of strength indicates a material's ability to resist "squeezing together?"
5. What type of strength indicates a material's ability to resist shock?
6. What type of strength indicates a material's ability to resist "pulling apart?"
7. What type of strength indicates a material's ability to resist repeated loading?
8. What type of strength indicates a material's ability to resist bending?
9. What type of strength indicates a material's ability to resist "sliding past?"
10. If a material refuses to stretch before it breaks, it is said to be _____.
11. Percent elongation is a good measure of the _____ of a material.
12. Name four important chemical properties of a metal.
13. What is dielectric strength?
14. A material that does not permit electricity to flow through it is said to have a high electrical _____.
15. What is thermal conductivity?
16. Which of the following materials has the greatest tensile strength?
    a. Steel.
    b. Wrought Iron.
    c. Aluminum.
    d. Zinc.
17. Which of the following materials has the best ability to resist compression?
    a. Brass.
    b. Aluminum.
    c. Cast Iron.
    d. Tin.
18. Which of the following materials is the heaviest?
    a. Brass.
    b. Copper.
    c. Bronze.
    d. Aluminum.
    e. Cast Iron.
19. Which of the following materials would conduct electricity the fastest?
    a. Copper.
    b. Aluminum.
    c. Steel.
    d. Nickel.
    e. Tungsten.
20. Which of the following materials has the highest melting temperature?
    a. Steel.
    b. Cast Iron.
    c. Tungsten.
    d. Zinc.
    e. Bronze.

7 CRYSTAL STRUCTURE

After studying this chapter, you will be able to:

☐ Explain how a crystal is formed in metal.
☐ Discuss what space lattices and dendrites are.
☐ Describe what atoms look like inside a crystal.
☐ Explain how temperature affects the growth of a crystal.
☐ Tell what is meant by "grain size."

CRYSTALLIZATION

Have you ever admired the crystalline pattern of a snowflake or frost on a window? Iron and steel form crystalline patterns in the same way that snow and ice do. Snow crystals and frost patterns start out as water vapor in the air. They change from liquid water vapor to solid crystals when they encounter heat. Crystalline patterns of frost form when water vapor condenses on a window pane that is warmer than the outside air.

Iron and steel can be heated so hot that they melt into a liquid. When this liquid gradually cools, CRYSTALS slowly begin to form as it solidifies. Tiny crystals form first. They keep growing until the crystals all stand tightly next to one another, "elbow to elbow." See Fig. 7-1.

Would you think that after these crystals solidify, their atoms would appear in a regular, precise formation or all mixed up? The answer may surprise you, Fig. 7-2. Their internal structure is very regular and precise. The atoms line up like a military band, marching in a parade down Main Street. Different crystals may not be in the same formation with respect to each other. However, the tiny atoms form long, neat rows in all directions within each crystal.

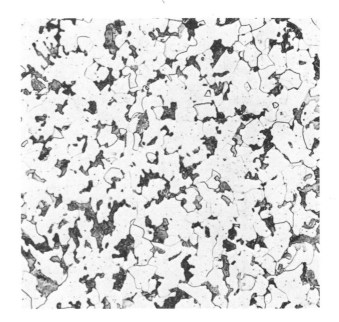

Fig. 7-1. Microphotography reveals this crystalline pattern of a 1020 steel ferrite structure with some pearlite. Magnification is 100x. (Buehler Ltd.)

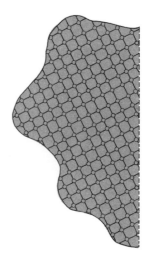

Fig. 7-2. Atoms in a crystal appear in a regular precise formation.

SPACE LATTICE

This regular pattern of neat rows of atoms is known as a space lattice. A SPACE LATTICE is defined as "the arrangement of the atoms in a crystal." Some examples are shown in Figs. 7-5, 7-7, 7-9, and 7-11.

Atoms in space lattices are so small that they are invisible. Until recently, scientists were unable to see them with their most powerful microscopes. Today, the newest "electron microscope" can show particles as small as ten billionths of an inch. This permits scientists to faintly see the atom. Research is being continued to develop (or invent) more powerful microscopes so that someday we will be able to see atoms close-up. Not all metals have the same style of space lattice pattern or the same "marching formation."

UNIT CELL

If you were to take a large space lattice of metal and break it down to its smallest fundamental arrangement, this most fundamental arrangement would be known as a UNIT CELL. A space lattice is simply a group of unit cells — perhaps billions — in which each unit cell is identical.

Since each type of metal has its own space lattice formation, there are many different basic unit cells. Some examples of these are shown in Fig. 7-3.

Four common types of unit cells that metals form are classified as:
1. Body centered cubic.
2. Face centered cubic.
3. Close packed hexagonal.
4. Body centered tetragonal.

There are many more formations that atoms take. However, these four are the most common types in the study of metallurgy.

BODY CENTERED CUBIC SPACE LATTICE

The unit cell of a BODY CENTERED CUBIC space lattice is diagrammed in Fig. 7-4. It consists of eight atoms in a square cube. In the center of these eight atoms is a ninth atom which completes the formation. A group of these body centered cubic unit cells would take the formation shown in Fig. 7-5.

It should be pointed out that the black lines shown between these atoms do not exist. They are shown in the diagram merely to help you visualize the space relationship between the atoms. Also, the atoms are considerably closer together than Fig. 7-4 suggests. The atoms depicted here as balls are packed so closely together that they just about touch.

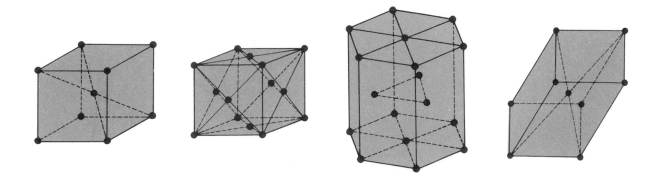

Fig. 7-3. Drawings depict unit cells for common types of space lattice formation.

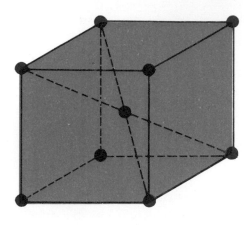

Fig. 7-4. Unit cell for a "body centered cubic" space lattice.

FACE CENTERED CUBIC SPACE LATTICE

The FACE CENTERED CUBIC space lattice is another very common unit cell found in metals. It also consists of eight atoms which form the corners of a cube. However, this formation does not have an atom in the middle of the cube. Instead, it has an atom in the middle of each of the six faces, Fig. 7-6. Thus, the total number of atoms in the basic unit cell of a face centered cubic space lattice is 14.

Metals that commonly take the body centered cubic formation, include chromium, molybdenum, tantalum, tungsten, vandium, columbium, and the FERRITE FORM OF IRON.

To help you visualize how these unit cells become part of a space lattice, see Fig. 7-5. Here, a series of body centered cubic unit cells are connected in a small space lattice. Next, visualize billions of these unit cells side by side instead of just the few shown in Fig. 7-5. Billions of these cells are required to make just one cubic inch of iron.

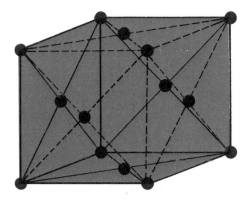

Fig. 7-6. Unit cell for a "face centered cubic" space lattice.

Metals that commonly take this formation are aluminum, copper, gold, lead, nickel, platinum, silver, and AUSTENITIC IRON.

A series of face centered cubic unit cells form the small space lattice shown in Fig. 7-7.

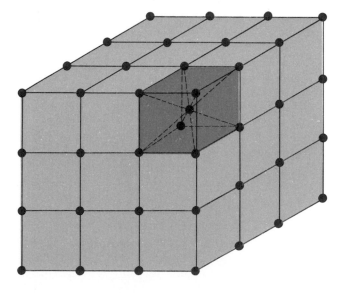

Fig. 7-5. A group of body centered cubic unit cells.

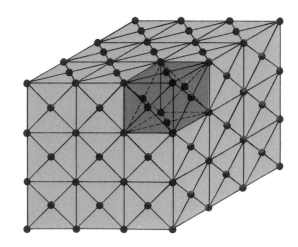

Fig. 7-7. A group of face centered cubic unit cells.

CLOSE PACKED
HEXAGONAL SPACE LATTICE

The CLOSE PACKED HEXAGONAL space lattice is a very brittle formation. It is found in metals that have little ductility or stretching ability. Its structure is quite different from the two space lattice structures, Figs. 7-5 and 7-7, we have studied thus far. Compare these to the close packed hexagonal unit cell shown in Fig. 7-8 and the space lattice in Fig. 7-9.

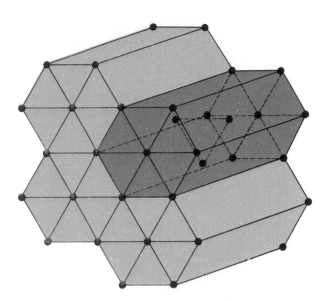

Fig. 7-9. A group of close packed hexagonal unit cells.

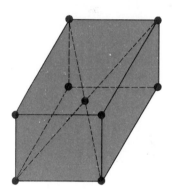

Fig. 7-10. Unit cell for a "body centered tetragonal" space lattice.

Fig. 7-8. Unit cell for a "close packed hexagonal" space lattice.

The unit cell consists of 17 total atoms, seven on each of two hexagonal shaped ends and three spaced in the center. See Fig. 7-8.

Metals that commonly take the close packed hexagonal space lattice formation include cadmium, cobalt, magnesium, titanium, beryllium, and zinc.

BODY CENTERED
TETRAGONAL SPACE LATTICE

The BODY CENTERED TETRAGONAL space lattice structure is almost identical to the body centered cubic. However, the faces of this structure are rectangular instead of square, Figs. 7-10 and 7-11. As with the body centered cubic, nine atoms make up the basic unit cell.

There is only one common metal that takes this formation. It is MARTENSITIC IRON.

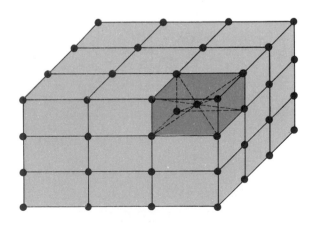
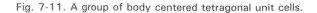

Fig. 7-11. A group of body centered tetragonal unit cells.

This is the hardest, strongest, and most brittle of the four space lattice structures we have discussed.

SPACE LATTICE STRUCTURES IN IRON AND STEEL

It may have surprised you to see that iron was listed three times when we discussed the four different unit cell types. Iron is unusual in that it can take three different space lattice structures. As iron goes through a temperature change, its atoms realign themselves into new geometric patterns. This has a great effect on the strength and hardness and ductility of the iron.

FERRITIC IRON or ferrite takes the body centered cubic lattice structure formation. FERRITE is basic iron at room temperature which has not previously been heat-treated.

AUSTENITIC IRON or austenite takes the face centered cubic lattice structure. AUSTENITE is the structure that iron takes at elevated temperatures. In other words, if ferrite is heated, it gradually becomes austenite when high temperature is reached. As it is becoming austenite, the atoms are reshuffling within the crystal. The atoms are realigning themselves into a new space lattice formation.

MARTENSITIC IRON or martensite has the body centered tetragonal crystal lattice structure. MARTENSITE is iron at room temperature that has previously been heated and suddenly quenched. The heating and quenching operation serves to produce this third different geometric pattern. Heating and sudden quenching tend to harden metal. Therefore, martensite is the strongest and hardest and most brittle of the three iron structures.

Fig. 7-12 summarizes the heat-treating processes which produce ferrite, austenite, and martensite.

While most metals tend toward one basic crystal lattice structure at all times, iron and steel are different. "Iron is not iron is not iron." The chart in Fig. 7-13 summarizes the characteristics of these three basic forms of iron.

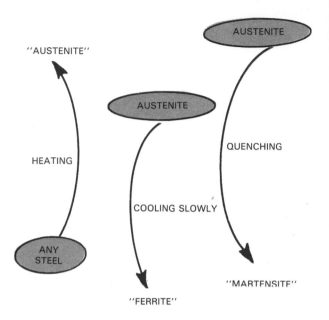

Fig. 7-12. The heat-treating processes that produce ferrite, austenite, and martensite are graphically shown.

FERRITE	AUSTENITE	MARTENSITE
Body Centered Cubic Lattice Formation	Face Centered Cubic Lattice Formation	Body Centered Tetragonal Lattice Formation
Exists At Low Temperature	Exists At High Temperature	Exists At Low Temperature
Magnetic	Nonmagnetic	Magnetic
Less Hardness Than Most Steels	Essentially No Hardness	More Hardness Than Most Steels
Less Strength Than Most Steels	Essentially No Strength	More Strength Than Most Steels
Ductile	(Not Applicable)	Brittle
Less Internal Stress Than Most Steels	(Not Applicable)	More Internal Stress Than Most Steels

Fig. 7-13. Chart compares characteristics of ferrite, austenite, and martensite.

TRANSFORMATION TEMPERATURE

As ferritic iron changes to austenite, there are two important temperatures to understand:

The LOWER TRANSFORMATION TEMPERATURE is the temperature at which the body centered cubic structure STARTS to change to the face centered cubic structure. It is the temperature at which ferrite STARTS to change to austenite.

The UPPER TRANSFORMATION TEMPERATURE is the temperature at which the body centered cubic lattice structure has COMPLETELY changed to face centered cubic. It is the temperature at which no ferrite exists. All of the iron structure above the upper transformation temperature is austenite. See Fig. 7-14.

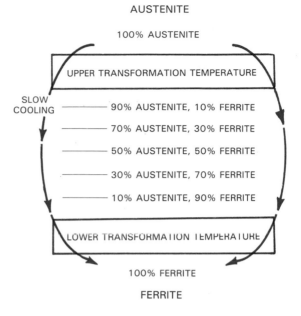

Fig. 7-15. Ferrite is transformed to austenite and back again in stages between the lower transformation temperature and the upper transformation temperature during slow cooling.

If the steel is quenched rapidly, the austenite changes to martensite (body centered tetragonal) instead of returning to ferrite, Fig. 7-16.

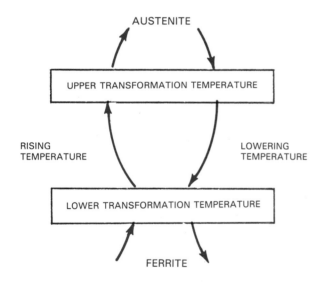

Fig. 7-14. Below the lower transformation temperature, ferrite exists without austenite. Above the upper transformation temperature, austenite exists without ferrite.

As austenite is cooled back down to room temperature, these two temperatures are important. When the UPPER transformation temperature is reached, austenite STARTS to change back to ferrite (body centered cubic). When it has cooled to the LOWER transformation, the structure becomes 100 percent ferrite (body centered cubic), Fig. 7-15.

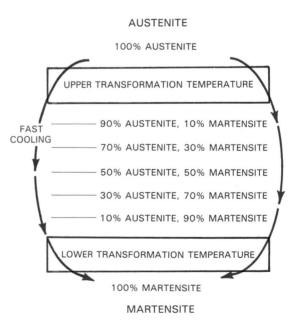

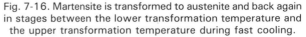

Fig. 7-16. Martensite is transformed to austenite and back again in stages between the lower transformation temperature and the upper transformation temperature during fast cooling.

The lower transformation temperature for all iron and steel is approximately 1330°F (721°C). The upper transformation temperature varies for each metal. It may be as low as 1330°F (721°C) or as high as 2000°F (1 094°C).

GROWTH OF A CRYSTAL

Imagine that some iron has been heated above the upper transformation temperature. Pretend, in fact, that is has been heated above the melting temperature. The iron has become liquid. If this iron is left at this elevated temperature, it will remain as a liquid molten metal. If it is cooled, it will begin to slowly solidify, unit cell by unit cell, until the entire mass of metal has solidified and become ferrite.

This solidifying process is interesting. Picture the iron in a molten state, Fig. 7-17. As the temperature is very slowly lowered, one spot in the mass of iron eventually will become cool enough to solidify. One unit cell will be formed at this point, Fig. 7-18. As the temperature continues to drop, more unit cells will be formed throughout the molten iron. Some of these newly solidified unit cells will attach themselves onto the first unit cell and it will tend to grow branches. See Fig. 7-19.

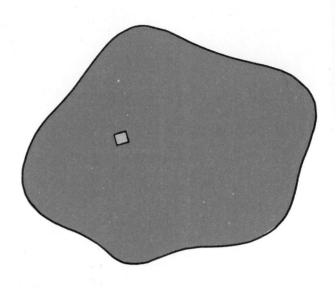

Fig. 7-18. As molten iron is cooled slowly, eventually one unit cell solidifies.

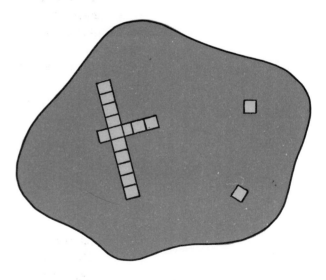

Fig. 7-19. As molten iron is cooled still further, branches begin to attach themselves onto the first unit cell.

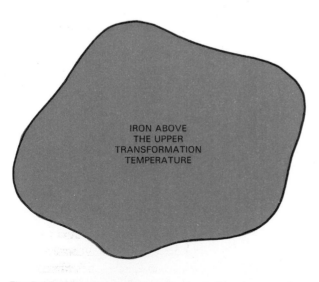

IRON ABOVE
THE UPPER
TRANSFORMATION
TEMPERATURE

Fig. 7-17. When iron is heated far above the upper transformation temperature, it is a molten state.

As these unit cells continue to solidify, they tend to collect in groups or in small colonies, Fig. 7-20. If the temperature is lowered very slowly, many of these colonies grow quite large. As the colonies grow, they grow larger and longer branches and resemble a skeleton. These sprouts or growing colonies, which resemble an unfinished frost pattern, are known as DENDRITES.

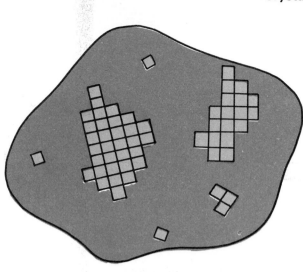

Fig. 7-20. As unit cells grow more and more branches, small colonies begin to develop which are known as dendrites.

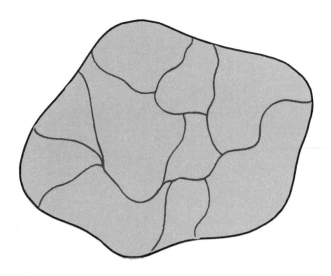

Fig. 7-22. Upon completion of the solidification of iron, there are boundaries between each colony group. Each colony within the boundary is a grain.

Growth of the dendrites continues. As solidification nears completion, the boundaries of some of these colonies will contact each other, Fig. 7-21. Here a conflict exists. The unit cell axis of each colony is different. The colonies tend to compete for the membership of the last few unsolidified cells.

When the entire mass of iron finally becomes solid, there will be boundaries between the colony groups. See Fig. 7-22. These boundaries are

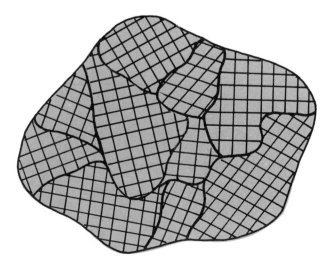

Fig. 7-21. As solidification nears completion, boundaries of these colonies contact each other.

noticeable under a microscope. In fact, in many materials these boundaries between colony groups are visible to the naked eye.

Each colony within the boundary is known as a grain. A GRAIN is any portion of a solid, which has external boundaries and an internal atomic lattice arrangement that is regular. A grain could also be called "a full grown dendrite."

The term CRYSTAL has several meanings. Most often, "crystal" means exactly the same thing as "grain." However, in a technical sense, a crystal is anything from a unit cell to a fully grown grain. Thus, a crystal may mean a grain, a unit cell, or a dendrite.

GRAIN SIZE VERSUS TIME

The size of the grain or crystal has a profound effect on strength, hardness, brittleness, and ductility.

If metal is cooled from the molten state very, very, slowly, the colonies have much time to add on members. Therefore, if the iron is cooled very slowly, these colonies will have time to grow larger and larger, and very large grain size will result.

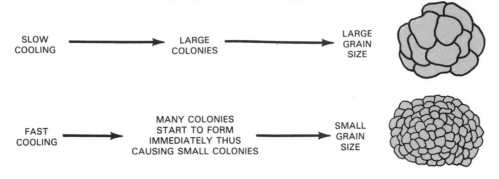

Fig. 7-23. Slow cooling produces large grain size. Fast cooling produces small grain size.

If, on the other hand, the material is cooled very rapidly, many more colonies will immediately start to spring up all over the crystal. Then, the size of each colony is limited because there are so many more colonies formed. Therefore, slow cooling produces a large grain size. Rapid cooling produces a small grain size. See Fig. 7-23.

EFFECT OF GRAIN SIZE

If you take a sheet of paper and start to tear it on one edge, the paper has little resistance to prevent the tear from moving across its entire surface. On the other hand, if a tear is started in a sheet formed by paper bits glued together, Fig. 7-24, it is more difficult for the tear to move smoothly and rapidly across the sheet.

Metals behave in the same manner. Those with large grain sizes are easier to tear or break or frac-

ture. Those with small grain size have high resistance to fracture. The small crack that starts has more difficulty moving across a series of small grains than across one large open field, Fig. 7-25.

In summary: the smaller the grain size, the greater the strength; the larger the grain size, the less strength. Therefore, in metallurgy, efforts are taken to keep the grain size as small as possible when strength is important.

Since strength, hardness, and brittleness are three inseparable partners, small grain size not only will yield better strength characteristics, but also will result in a harder material and a more brittle material. Therefore, if it is important for a material to have strength, wise metallurgists will

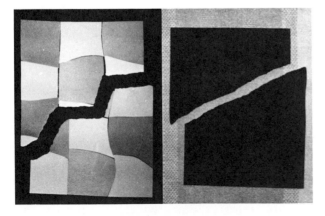

Fig. 7-24. Paper is more difficult to tear when it is made up of many small bits glued together.

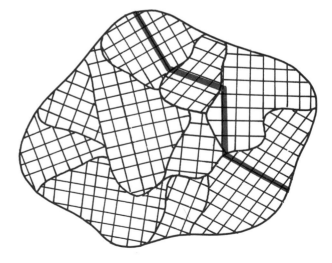

Fig. 7-25. Cracks have more difficulty moving across a series of small grains than across large grains.

attempt to obtain small grain size. If ductility is more important than strength, they will want to attain large grain size.

GRAIN SIZE—COOLING SPEED— STRUCTURE—PROPERTIES

Fig. 7-26 summarizes the results of large and small grain size. Note that the ferrite lattice structure has the advantage of being more ductile. Therefore, ferrite is more easily machined and less prone to cracking.

The martensite structure, on the other hand, gives greater strength and hardness. Martensite, however, is more prone to brittleness, cracking, and distortion because of the fast cooling that is required to attain the small grain size.

	FERRITE	MARTENSITE
Grain Size	Large	Small
Cooling Speed	Slow	Rapid
Strength	Lower	Higher
Hardness	Lower	Higher
Ductility	Ductile	Brittle
Distortion & Cracking	Little Tendency	Greater Tendency
Machinability & Formability	Best	Difficult

Fig. 7-26. Chart compares two types of room temperature iron structures, ferrite and martensite.

TEST YOUR KNOWLEDGE

Write your answers on a separate sheet of paper. Do not write in this book.

1. The arrangement of atoms in a crystal is very orderly. The most fundamental arrangement that shows the basic structure of the most simple space lattice is known as a _____.

2. As a substance is cooled and the material begins to solidify, the branches that start to grow are known as _____.
3. What has external boundaries and a regular internal lattice structure?
4. How many atoms are in the unit cell of a "body centered cubic" lattice structure?
5. As a large piece of material started to cool, suddenly there were unit cells. After the material continued to cool to room temperature, you were finally able to see the _____.
6. One of the four lattice formations is not typical of any iron structure. It is _____.
7. How many atoms are in the unit cell of a "face centered cubic?"
8. Another name for a grain is a _____.
9. Ferritic iron or ferrite takes which lattice structure formation?
   a. Body centered cubic.
   b. Face centered cubic.
   c. Body centered tetragonal.
   d. Close packed hexagonal.
10. Austenitic iron or austenite takes which lattice structure formation?
    a. Body centered cubic.
    b. Face centered cubic.
    c. Body centered tetragonal.
    d. Close packed hexagonal.
11. Martensitic iron or martensite takes which lattice structure formation?
    a. Body centered cubic.
    b. Face centered cubic.
    c. Body centered tetragonal.
    d. Close packed hexagonal.
12. Slow cooling will cause which of the following:
    a. Small grain size.
    b. Medium grain size.
    c. Large grain size.
13. Fast cooling will produce which of the following:
    a. Ferrite.
    b. Austenite.
    c. Martensite.

FAILURE AND DEFORMATION OF METAL

After studying this chapter, you will be able to:

☐ Explain what actually happens inside a piece of metal when it breaks.
☐ State what is meant by deformation of metal.
☐ Describe the different types of metal failure or breakage.
☐ Define the term "work hardening" and tell why it is important.

DEFORMATION

DEFORMATION occurs when metal is willing to stretch. It is the amount that a material increases or decreases in length when it is loaded (force is applied). Materials like cast iron and concrete and peanut brittle are very unwilling to stretch. They never will have much deformation. Some other materials like aluminum and polyethylene and rubber stretch far more. They have a high degree of deformation.

Before a metal fractures, it may stretch a lot or it may stretch very little. The material represented in Fig. 8-1 breaks when the force on

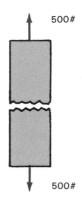

Fig. 8-1. This material does not stretch before breaking.

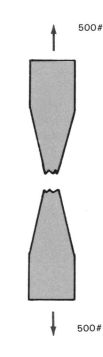

Fig. 8-2. This material stretches considerably before breaking.

it reaches 500 pounds. Before it breaks, the metal does not stretch at all.

By contrast, the metal of Fig. 8-2 also fails when the force reaches 500 lbs. However, before it fails, it stretches considerably. Thus, the amount of deformation that a material goes through before failure does not determine the total amount of force that it is able to take before it breaks.

Fig. 8-3 shows several materials before and after they fractured under test. Aluminum, low carbon steel, nylon, and polyethylene all are considered to be DUCTILE materials. They stretch considerably before failure. Cast iron is a BRITTLE material because it does not stretch before it breaks.

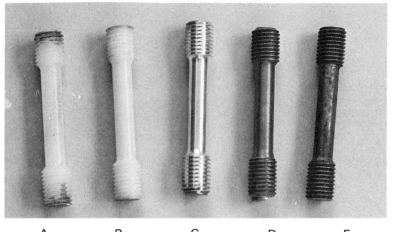

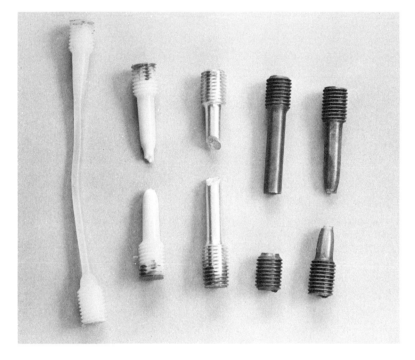

Fig. 8-3. These five samples are the same length, diameter, and general configuration before being tested: A — Polyethylene. B — Nylon. C — Aluminum. D — Cast iron. E — Steel. After fracture, some materials show greater ductility than others. Polyethylene shows the greatest ductility. Cast iron shows the least ductility or greatest brittleness.

BRITTLENESS—DUCTILITY— CLEAVAGE—SHEAR

A material that stretches much before it fails is said to be "ductile," regardless of how much force it resists before failure (refer to Chapter 6). A material that stretches very little before it fails is said to be "brittle," regardless of how much force it takes before failure.

Brittle materials fail in CLEAVAGE, Fig. 8-4. Ductile materials fail in SHEAR, Fig. 8-5.

CLEAVAGE—BRITTLE FAILURES

A material that failed in cleavage is shown in Fig. 8-6. Cast iron is a typical example of this type of failure. In a brittle failure, the atoms merely break apart from each other and separate.

BRITTLE FAILURE
1. Fails in cleavage
2. Negligible stretch before failure
3. Examples: a. Cast iron b. Concrete c. Wood

Fig. 8-4. Chart gives characteristics of a brittle failure.

DUCTILE FAILURE
1. Fails in shear
2. Much stretching before failure
3. Examples: a. Aluminum b. Low carbon steel c. Rubber

Fig. 8-5. Chart shows characteristics of a ductile failure.

Fig. 8-6. This brittle cast iron sample failed in cleavage.

A cleavage failure or a brittle failure appears bright, rough, and granular, Fig. 8-7.

Occasionally, ductile materials may exhibit a "brittle" failure. They fail in this way when they are attacked by rapid shock loads and do not have time to begin to stretch.

Fig. 8-7. A brittle failure appears bright, rough, and granular.

SHEAR—DUCTILE FAILURE

When a ductile material fails, the atoms slide past each other within the crystals. This causes the material to stretch, Fig. 8-8. This shearing

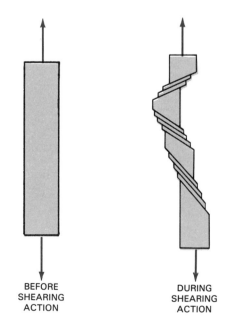

BEFORE
SHEARING
ACTION

DURING
SHEARING
ACTION

Fig. 8-8. Internal shearing action causes a ductile material to stretch.

134

type action will sometimes cause materials to "neck down" or become thinner and longer near the failure point. See Fig. 8-9.

Shear failures appear dull, smooth, velvety, and fibrous.

Shear failures take one of two forms, either "slip" failures or "twinning" failures.

Fig. 8-10. Aluminum is a ductile material that fails by shear.

Fig. 8-9. Both steel and nylon show the tendency to neck down before failure.

SLIP FAILURES

Fig. 8-10 shows a ductile material that fails by shear. Fig. 8-11 shows the action of the atoms during this shear type of failure known as SLIP. Instead of dividing in the middle of a crystal, the atoms tend to slide past each other and move down one row at a time. As they slide past each other, one row at a time, the part deforms and the part becomes longer and longer, thinner and thinner. Eventually, a break may occur.

Slip takes place along certain crystal lines which are called SLIP PLANES. An entire block of atoms moves over another entire block of atoms. This action takes place through several thousand atomic layers. The movement moves forward one step at a time, Fig. 8-12.

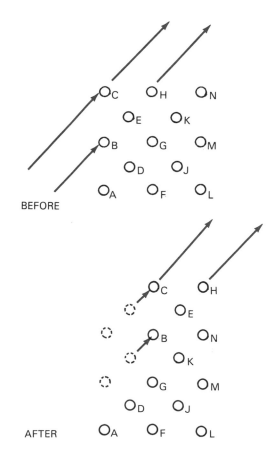

Fig. 8-11. In a slip failure, atoms slide past each other.

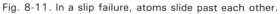

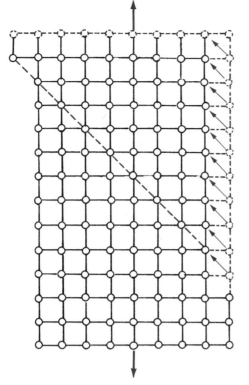

Fig. 8-12. A slip failure moves forward one step at a time.

Fig. 8-13 shows a body centered cubic structure that has slipped two atomic layers.

As a material stretches, several slip planes may be in action at the same time. This group of slip planes is called a SLIP BAND, Fig. 8-14. When many slip planes are present, deformation is relatively easy. When fewer slip planes are involved, slip is more difficult. The metal is stronger but less ductile.

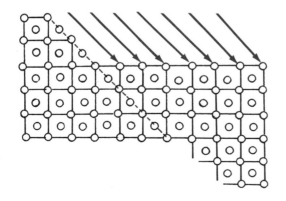

Fig. 8-13. This body centered cubic structure has slipped two atomic layers.

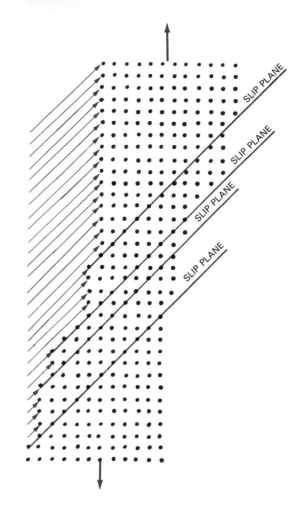

Fig. 8-14. Several slip planes may occur during slip failure and are called a slip band.

TWINNING

The action of TWINNING is very similar to the action of slip. Both are ductile failures. Both fail by shear. Some metals deform easier by twinning than by slip. Some metals deform by both slip and twinning.

In twinning, a zone within the crystal lattice structure is deformed from its original space lattice formation. The two lines that separate the deformed zone from its parents are known as MIRROR LINES or TWINNING PLANES. See Fig. 8-15.

The formation of atoms on either side of the mirror line or twinning plane is the same. If a

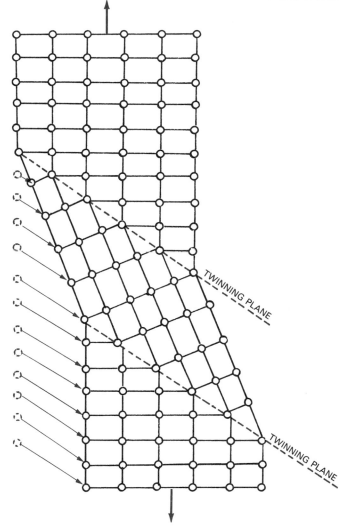

Fig. 8-15. A twinning failure involves two mirror lines or twinning planes.

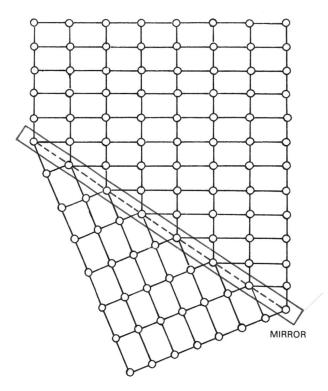

Fig. 8-16. In a twinning failure, the image on either side of the mirror line is identical.

LARGE AND SMALL CRYSTALS

Which material will be able to resist more force, one with large crystals or small crystals? In Chapter 7, we learned that a material with small crystals is the strongest. Small crystals resist cracking better. Slip and twinning help to clarify why this is true.

If slip or twinning occurs, the slip plane can move rapidly all the way across a crystal. However, when the slip plane reaches the end of a crystal, and has to continue across a second metal crystal, fracture or additional slip becomes more difficult, Fig. 8-17. The second crystal will have a different direction of lattice orientation, so the crack must change direction. Failure is made more difficult each time a crystal boundary is reached. Therefore, a material that has many crystals which are small is generally able to resist fracture much better than a material that has only one crystal or a few crystals, Fig. 8-18.

mirror were placed perpendicular to the twinning plane, and on the twinning plane, Fig. 8-16, the image in the mirror would be identical to the orientation of the group of atoms behind the mirror.

Twinning causes a new lattice orientation in the twinning region. This twinning region may involve millions of atoms. Essentially a long block or plane of atoms is affected. All the atoms take the sliding action simultaneously, similar to the way that slip takes place.

After twinning progresses far enough, a separation failure may also take place.

This emphasizes the importance of rapid cooling of metal. With rapid cooling, small crystals are formed. The small crystals resist failure much better than large crystals which are formed by slow cooling.

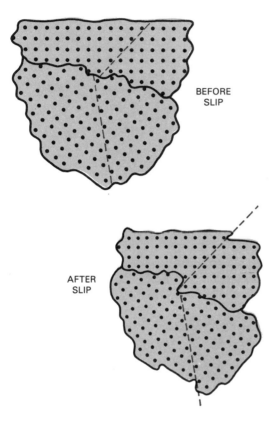

Fig. 8-17. Slip becomes more difficult when the slip plane reaches the end of a crystal.

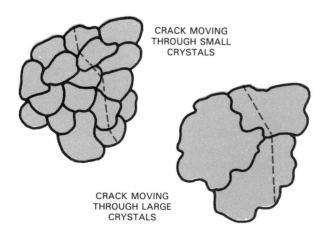

Fig. 8-18. Small crystals can resist fracture better than large crystals.

WORK HARDENING

As materials are stretched out of their normal shape, a strange phenomenon known as WORK HARDENING or STRAIN HARDENING occurs.

Consider two machine parts. One part is attacked by a force. Before it breaks, the force is removed and the part relaxes. The force is again applied, then removed before failure takes place. *If this application of force and removal of force is repeated over and over, do you think this material will eventually fail at a lower force level than a second machine part which is loaded until failure the first time? In other words, does the repeated application of force and removal of force weaken a material or strengthen it?*

We need to analyze what happens. A material that receives a few preliminary force applications WORK HARDENS. It eventually becomes stronger and harder (although it also becomes more brittle). Thus, in Fig. 8-19, the first (work-hardened) machine part (A1) will break at a higher force value than the second machine part (A2) that was loaded until failure the first time.

The reason that this work-hardening phenomenon occurs is not totally understood. However, it is assumed that as the atoms are forced into a stretched position, they tend to lock into a stronger formation than they originally had. (This is similar to the way weight lifters readjust their feet each time a larger weight is about to be lifted.) As stretching takes place, some atoms are torn from their original lattice position and move to a new spot in between slip planes. Then, they become a roadblock for the sliding of one plane over another. Thus, the strength and hardness increase but some ductility and elasticity is lost.

In manufacturing, this work-hardening phenomenon is often a blessing. In processes such as forging, extruding, drawing, and rolling, "cold working" takes place as the metal is repeatedly compressed. Added strength that occurs because of this work hardening is a free benefit.

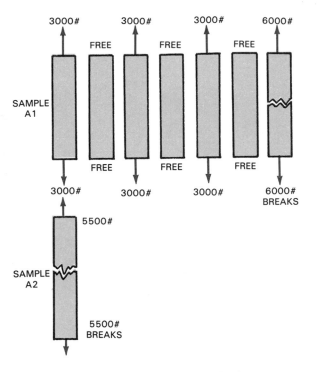

SAMPLE A1

3000#   3000#   3000#   6000#
FREE   FREE   FREE
FREE   FREE   FREE
3000#   3000#   3000#   6000#
                        BREAKS

SAMPLE A2

5500#

5500#
BREAKS

Fig. 8-19. Work hardening causes a material to become stronger and harder and more brittle.

TEST YOUR KNOWLEDGE

Write your answers on a separate sheet of paper. Do not write in this book.

1. Name two metals that show a high degree of deformation or ductility.
2. Name one metal that shows a very low degree of deformation.
3. If a metal stretches before it breaks, is it more likely to fail in shear or cleavage?
4. Under what circumstances will a ductile material fail in cleavage instead of in shear?
5. If the atoms inside a crystal tend to slide past each other, one row at a time, the failure that is occuring is known as:
   a. Twinning.
   b. Slip.
   c. Cleavage.
6. A shear type failure generally occurs after:
   a. Twinning.
   b. Slip.
   c. Either.
   d. Neither.
7. Which is generally stronger, a material with small fine crystals or a material with large crystals?
8. Which is generally able to resist cracking better, a material with small, fine crystals or a material with large crystals?
9. Rapid cooling or quenching causes:
   a. Small crystals.
   b. Medium crystals.
   c. Large crystals.
10. If a force is applied and removed over and over again, the material will begin to _____ _____.
11. In forging and drawing, the material becomes stronger automatically, due to a phenomenon called _____ _____.

IRON-CARBON DIAGRAM

After studying this chapter, you will be able to:

☐ Describe five important structural forms of steel and iron.
☐ Interpret an "iron-carbon diagram."
☐ Use this diagram to determine the temperatures to which steel must be heated to cause it to harden.

FERRITE, CEMENTITE, AND PEARLITE

Steel is iron with more than 0 percent carbon but less than approximately 2.0 percent. See chart in Fig. 9-1. However, steel behaves very differently as the amount of carbon increases from 0 percent to 2.0 percent.

Steel with very little carbon in it is called FERRITE (shown at left end of chart in Fig. 9-1). Steel that has a carbon content above 0.8 percent contains some CEMENTITE (shown at right end of steel region in chart). Steel that has approximately 0.8 percent carbon is called PEARLITE.

All three of these forms — ferrite, cementite, and pearlite — represent steel at room temperature.

FERRITE

FERRITE is almost pure iron. It has little "desire" to dissolve carbon. Therefore, little carbon is in it. Since carbon gives steel the ability to become strong and hard, ferrite is a very weak steel. Ferrite exists at low temperatures only, and it is magnetic.

CEMENTITE

CEMENTITE is actually a compound of iron and carbon, known as "iron carbide." Its chemical formula is Fe_3C. Cementite contains 6.67 percent carbon by weight. However, cementite is present in the alloy between 0.8 and 6.67 percent carbon. As the percentage of carbon increases, more and more cementite is present, until at 6.67 percent carbon, the entire mixture is cementite. Below approximately 2.0 percent carbon, the alloy is still considered to be steel. Above that percentage, it becomes cast iron. (See Chapter 3.)

After heat-treating, cementite can become very strong and hard. It also exists at room temperature, and it is magnetic.

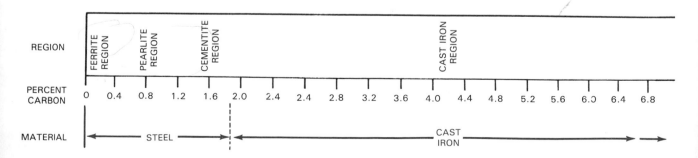

Fig. 9-1. Steel contains approximately 0 to 2.0 percent carbon. Cast iron contains more than 2.0 percent carbon.

PEARLITE

PEARLITE is a solid solution. It is a mixture of ferrite and cementite. Fig. 9-2 shows a microscopic view of pearlite. Notice that under the microscope, pearlite appears as layers. It reminds one of an aerial view of many newly plowed fields. The black ridges are cementite and the white ridges are ferrite. Thus, pearlite is made up of layers of ferrite and layers of cementite adjacent to one another.

When we have approximately 0.8 percent carbon, the ferrite and the cementite balance suffi-

ciently in the making of pearlite, so that the entire microscopic sketch appears as ridges. If there is less than 0.8 percent carbon, there will be a mixture of ferrite and pearlite, and only part of the picture will appear to be ridges. If there is an excess of carbon over 0.8 percent, there will be a mixture of cementite and pearlite.

Pearlite exists at room temperature and is magnetic.

Fig. 9-3 shows the relationship of the composition of steel at various percentages of carbon. Ferrite is the form of steel at very low carbon content. Pure pearlite occurs at approximately 0.8 percent carbon. Between these two extremes, we have a mixture of ferrite and pearlite.

A mixture of cementite and pearlite occurs above 0.8 percent and extends all the way into the cast iron region.

AUSTENITE

AUSTENITE is another structural form of steel. It occurs at elevated temperatures. It is not magnetic. NOTE: You "met" austenite briefly in Chapter 7.

As the steel is heated above room temperature

Fig. 9-2. Microscopic view reveals details of a 1095 steel pearlite structure. Magnification is 500x. (Buehler Ltd.)

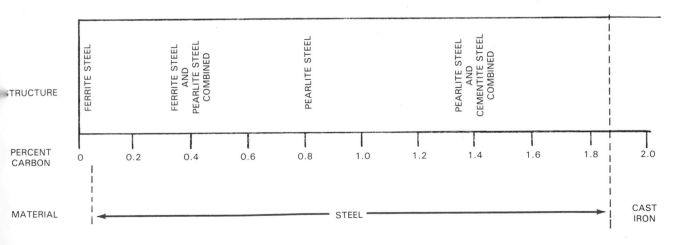

Fig. 9-3. Chart shows effect of percentage of carbon on presence of ferrite, pearlite, and cementite in steel.

(to an elevated temperature where it becomes austenite), its structure changes from body centered cubic to face centered cubic. See Fig. 9-4. If it is cooled down slowly back to room temperature, face centered cubic changes back again to body centered cubic and becomes ferrite, pearlite, and/or cementite.

IRON-CARBON PHASE DIAGRAM

A simplified version of the IRON-CARBON PHASE DIAGRAM is shown in Fig. 9-5. An industrial version of the diagram is pictured in Fig. 9-6. From the iron carbon phase diagram you can tell what structure iron takes at any given temperature — if you know the percent carbon in the steel. From the diagram, you can tell if you have ferrite, pearlite, cementite, austenite or any combination of the four. But, first you must know:

1. The temperature of the steel.
2. The percent carbon it contains.
3. Its past heat-treating history.

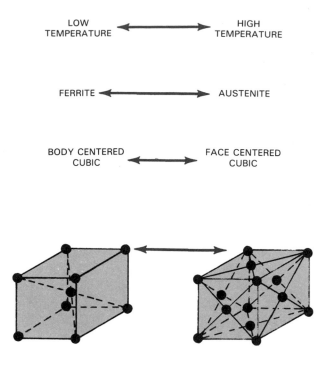

Fig. 9-4. Diagram compares structural differences in low temperature and high temperature steel.

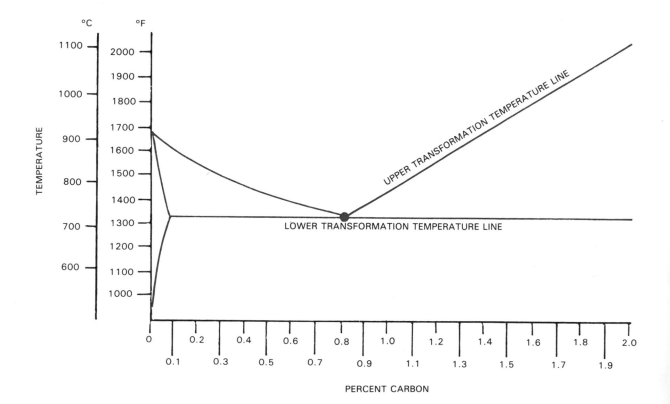

Fig. 9-5. Basic iron-carbon phase diagram gives steel temperature scale on vertical axis and percent carbon on horizontal axis.

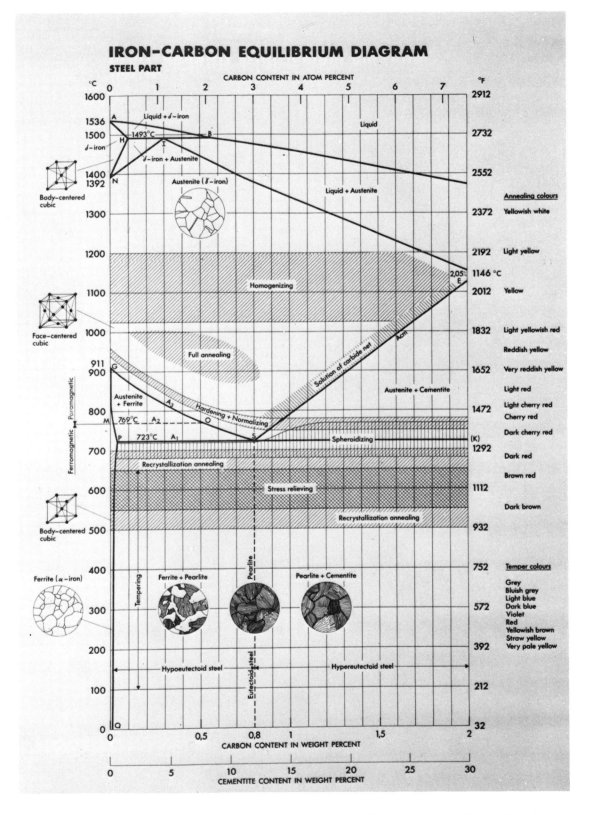

Fig. 9-6. This chart is industrial version of Fig. 9-5. It shows iron-carbon phase diagram extended to include higher temperatures. Do you recognize crystal structures shown at left? (Struers Scientific Instruments)

There are two key lines in the iron-carbon phase diagram. The lower of the two lines is the LOWER TRANSFORMATION TEMPERATURE LINE. The upper of the two lines is the UPPER TRANSFORMATION TEMPERATURE LINE. NOTE: Both of these transformation temperatures were discussed in detail in Chapter 7.

From these lines:
1. You know the temperature that iron starts transforming to austenite.
2. You know the temperature at which iron completes its transformation to austenite.

Any steel structure that occurs above the upper transformation temperature line is pure austenite, Fig. 9-7. Any steel structure that occurs below the lower transformation temperature line contains no austenite. The iron structure below the lower transformation temperature is made up of a combination of ferrite, pearlite, and cementite. See Fig. 9-8.

In the two triangular areas, between the upper and lower transformation temperature lines, there is a mixture of austenite with either ferrite or cementite. See Fig. 9-9. In the figure, also note the small triangular area at the left. This is the 100 percent ferrite region. In this area, all the carbon is totally dissolved in the iron. There is no pearlite or cementite.

USE OF AN IRON-CARBON DIAGRAM

One of the best ways to learn to use an iron-carbon phase diagram is to take a piece of steel and follow its journey through the diagram.

1. First assume that the steel is at room temperature and contains 0.4 percent carbon. This would put it at point A in Fig. 9-10. Since it is below the lower transformation temperature, this material contains no austenite. Point A is nearly midway between the pure pearlite line (at 0.8 percent carbon) and the pure ferrite limiting line. The material

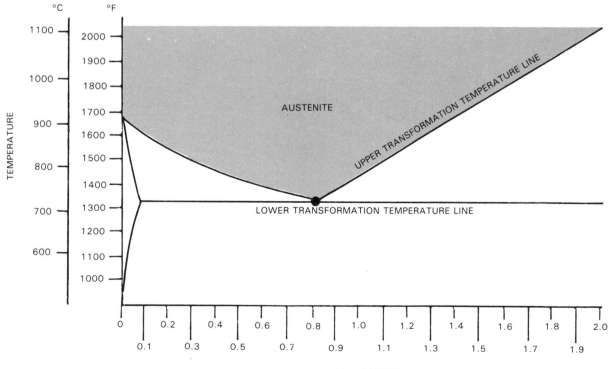

Fig. 9-7. Above upper transformation temperature line, steel takes the structure of austenite.

Iron-Carbon Diagram

is half ferrite and half pearlite.

2. Next assume that the material is heated to about 1000°F (538°C), at point B. What structure is it now? Austenite? Ferrite? Pearlite?

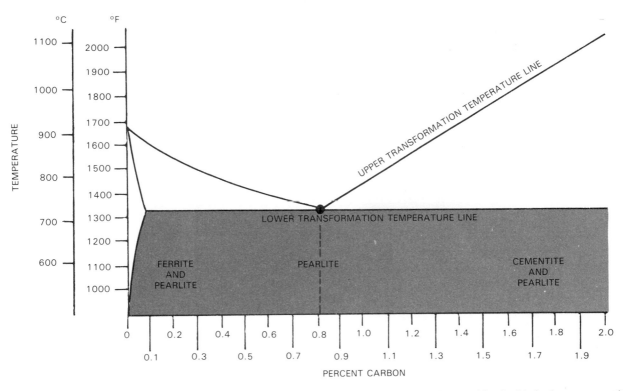

Fig. 9-8. Below lower transformation temperature line, steel takes the structure of pearlite combined with ferrite or cementite.

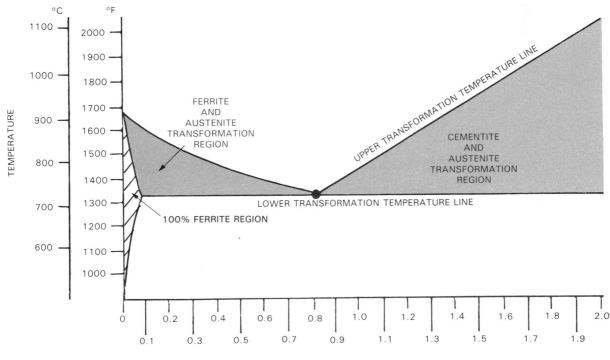

Fig. 9-9. Between the upper and lower transformation temperature lines, there is a mixture of austenite with ferrite or cementite.

Metallurgy Fundamentals

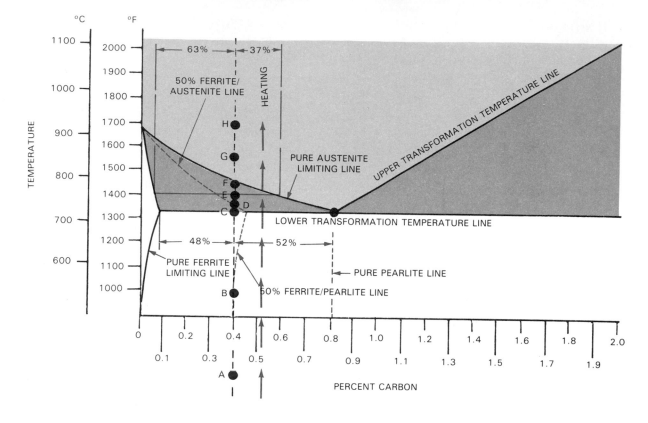

Fig. 9-10. The temperature journey of 0.4 percent carbon steel is shown on an iron-carbon phase diagram, during heating.

Cementite? It still is below the lower transformation temperature line. Therefore, no austenite is present. Point B is slightly closer to the pure ferrite limiting line than to the pure pearlite line since the pure ferrite line has moved slightly to the right. Its structure is between 50-51 percent ferrite and 49-50 percent pearlite.

3. Now assume that the material is heated to 1330°F (721°C), at point C. At 1330°F, the structure is 52 percent ferrite and 48 percent pearlite. (Point C falls just to the left of the 50 percent ferrite/pearlite line.) However, at this temperature, some key changes start to take place. Above 1330°F, the pearlite changes to austenite. Theoretically, all of the pearlite suddenly changes to austenite at a temperature near 1330°F. In reality, the change occurs over a small range in the vicinity of 1330°F—perhaps, from 1310°F to 1350°F.

4. The material is now heated to 1360°F (738°C), at point D. All of the pearlite has changed to

austenite. The structure is now 52 percent ferrite and 48 percent austenite.

5. At 1400°F (760°C), point E lies closer to the pure austenite limiting line than to the pure ferrite limiting line. Therefore, the material now contains more austenite than ferrite— approximately 63 percent austenite and 37 percent ferrite.

6. Just before the steel reaches the upper transformation temperature line, at point F, it has almost completely changed to austenite. Perhaps 10 percent ferrite is left, and the rest of the material has changed to austenite. Thus, the percentage now is 90 percent austenite and 10 percent ferrite.

7. At 1550°F (843°C), the material is completely changed to austenite. No ferrite or pearlite remains at point G.

8. As the material is heated higher to 1700°F (927°C) at point H, there is no further basic structural change. All of the steel has already changed to austenite. It will remain austenite.

Now refer to Fig. 9-11. If the material is cooled slowly, from 1700°F to 1550° to 1400° and all the way down to 1000°, it will change back to ferrite and pearlite in the reverse manner of the change from ferrite-pearlite to austenite in Fig. 9-10. Thus, at 1550°F, we still have all austenite. At 1400°F, we have 63 percent austenite and 37 percent ferrite. At 1360°F, we only have about 48 percent austenite and 52 percent ferrite.

At 1000°F, the entire steel structure has changed back to ferrite-pearlite and no austenite remains in the steel.

TRANSFORMATION TO MARTENSITE

If steel is quenched rapidly, Fig. 9-12, instead of cooled slowly, it changes to MARTENSITE instead of ferrite, pearlite, or cementite. NOTE: You also met martensite in Chapter 7.

At point H, we have 100 percent austenite. At point G, 1550°F, we have 100 percent austenite. At 1400°F, we have 37 percent martensite and 63 percent austenite. At 1000°F, 100 percent martensite exists.

EXAMPLES OF STEEL STRUCTURES

Fig. 9-13 shows a typical iron-carbon phase diagram with several steel lattice structures represented. The 10 points shown are at a variety of temperatures, and they indicate steels with various percentages of carbon.

Can you identify the various structures that correspond to each of the 10 letters shown in Fig. 9-13? Which are austenite? Which are ferrite? Which are pearlite? Which are cementite? Which are combinations of the four?

The steel lattice structures formed at points A and B both are 100 percent austenite because they are above the upper transformation temperature line.

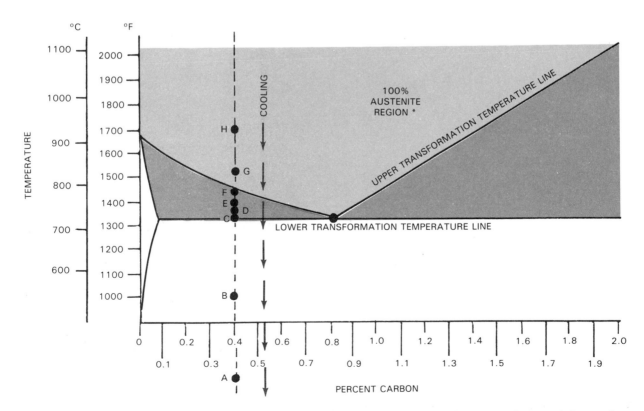

Fig. 9-11. The temperature journey of 0.4 percent carbon steel is traced on an iron-carbon phase diagram, during cooling.

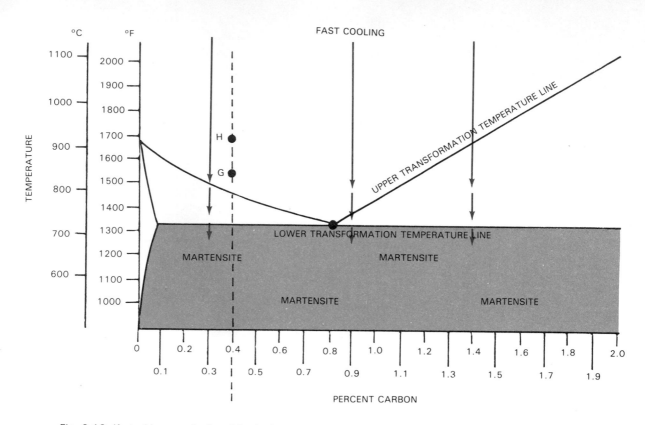

Fig. 9-12. If steel is quenched rapidly, it changes to martensite instead of ferrite, pearlite, or cementite.

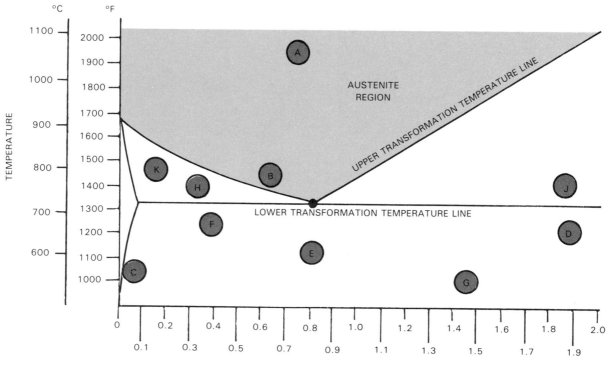

Fig. 9-13. Ten examples of steel structures are represented on an iron-carbon phase diagram.

Points C, D, and E, F, and G contain no austenite because they are below the lower transformation temperature line.

Since C is far to the left and has very little carbon, it is almost 100 percent ferrite. A small percentage of pearlite is also present.

Point E is located on the 0.8 percent carbon line, which is the home of 100 percent pearlite. Point F is halfway between pure pearlite and pure ferrite. Therefore, it is made up of approximately 50 percent ferrite and 50 percent pearlite. Points D and G are both located partway between the 100 percent pearlite line (at 0.8 percent carbon) and the 100 percent cementite line (at 6.7 percent carbon.) G is closer to the 100 percent pearlite line than D, but both are closer to the 0.8 percent than the 6.67 percent carbon location. Point D contains 81 percent pearlite and 19 percent cementite. Point G has an 89:11 ratio of pearlite to cementite.

Point H is in the transformation region; therefore, theoretically, all of its pearlite has transformed to austenite. Since it is midway between the pure ferrite region and the pure austenite region, its ratio of ferrite to austenite is 50:50.

Another point that is in the ferrite/austenite region is point K. However, point K lies twice as close to the pure ferrite line so it contains about 67 percent ferrite and 33 percent austenite.

Point J lies in the cementite/austenite region of the transformation portion of the figure. There is no pearlite since it lies above the 1330°F line. It is located closer to pure austenite (0.8 percent carbon) than to the 6.67 percent pure cementite line. Its structure can be calculated to be 81 percent austenite and 19 percent cementite. This calculation would be essentially identical to the calculation that gave a 81:19 ratio for point D.

It may be fun to make up additional points and try to figure out what steel structures they are.

Points C, D, E, F, G, H, J, and K could have been martensite instead of ferrite, pearlite, and cementite if these samples of steel had received a rapid cooling or quenching action. Heating them above the upper transformation temperature line, then quenching rapidly would have given them a martensitic structure. Then C, D, E, F, or G would have become 100 percent martensite. Point K would have been 33 percent austenite and 67 percent martensite. Point H would contain 50 percent austenite and 50 percent martensite. Point J would have 19 percent martensite and 81 percent austenite.

ADDITIONAL TERMS

There are some additional terms involving the iron-carbon phase diagram, Fig. 9-14, that are frequently used throughout industry:

EUTECTOID POINT refers to the point where the upper transformation temperature line, the lower transformation temperature line, and the 0.8 percent carbon pearlitic line all come together at one "grand highway intersection."

TRANSFER TEMPERATURE RANGE is the region between the upper and lower transformation temperature lines where the action is. In these two triangular areas, either austenite is changing to one of the low temperature structures or ferrite, pearlite, cementite, or martensite are changing to austenite.

HYPOEUTECTOID REGION is the region to the left of the 0.8 percent carbon line, Fig. 9-15. Any steel that falls in this region is known as hypoeutectoid steel.

HYPEREUTECTOID REGION is the region to the right of the 0.8 percent carbon line, Fig. 9-15. Any steel that falls in this region is known as hypereutectoid steel.

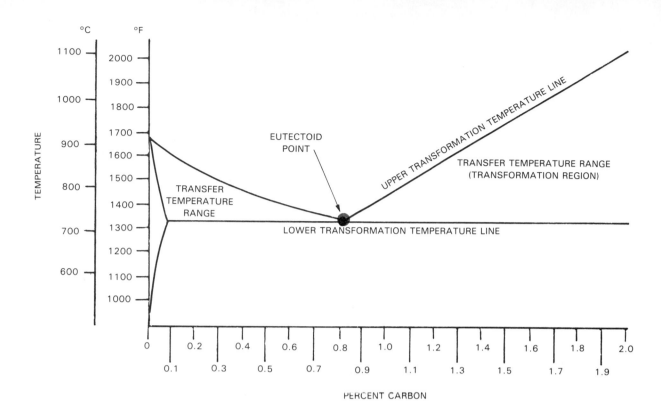

Fig. 9-14. Iron-carbon phase diagram shows eutectoid point, upper and lower transformation temperature lines, and transfer temperature range.

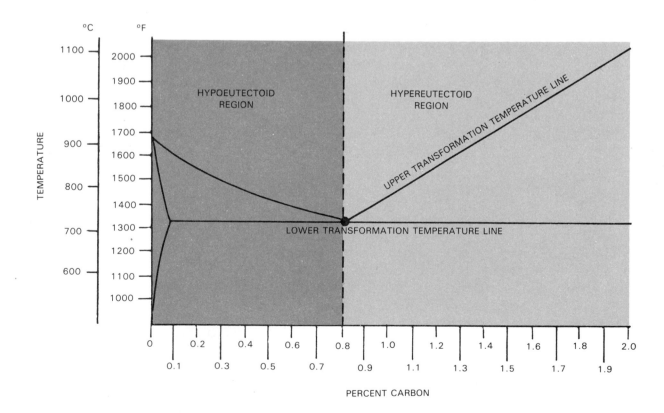

Fig. 9-15. Iron-carbon phase diagram is divided into hypoeutectoid and hypereutectoid regions.

QUENCHING refers to extremely rapid cooling. Generally, when the term is used, it refers to steel that has been heated to the austenite range, then cooled rapidly. Usually, this will result in a martensitic structure, rather than ferrite, pearlite, or cementite. Quenching methods include water quenching, oil quenching, air quenching, and other mediums that are described in Chapter 11.

SLOW COOLING, as the name implies, is the opposite of quenching. When a material is slow cooled from the austenite range, it will have changed to ferrite, cementite, pearlite, or a combination of the three. See Fig. 9-16.

Each material has its own limit as to how fast the quenching process must be, in order to obtain strong steel. Often, a part that is cooled "medium slowly" will end up as a combination of martensite and pearlite or one of the other two low temperature steel structures.

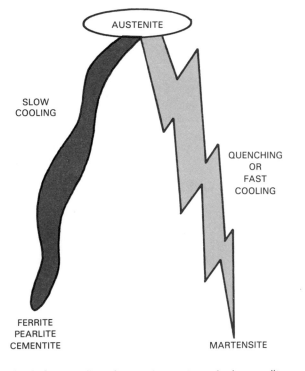

Fig. 9-16. Slow cooling of austenite produces ferrite, pearlite, or cementite. Quenching or fast cooling of austenite produces martensite.

HARDNESS—STRENGTH—BRITTLENESS—DUCTILITY—GRAIN SIZE

The hardness, strength, brittleness, ductility, and grain size of a material is affected greatly by the heating and cooling method.

If a material is heated to the austenitic range and then quenched very rapidly, martensite will generally occur. It will be a hard, strong material and have small grain size. However, it will be brittle.

If a material is heated to the austenitic range, and then cooled very slowly, it will change to ferrite, pearlite, or cementite. This structure is comparatively softer, less strong, more ductile, and would have larger grain size.

Hardness, strength, ductility, and small grain size are generally considered to be assets for a material (See Chapter 6). It would be ideal if one could heat and quench a material in such a manner that hardness and strength could be obtained without also getting brittleness.

Steel users generally have to make a choice among three situations:

1. If the material must be HARD AND STRONG, they will select a material that has been quenched rapidly. However, this material will usually be brittle and lack ductility.
2. If they need GREAT DUCTILITY in a material, it will be cooled slower and it will be very machinable and formable. However, it will not have good strength or hardness qualities.
3. If the steel user must have both STRENGTH AND DUCTILITY in a material, special alloys can be added to accomplish this. The addition of these alloys will increase the cost of the material. Often, however, higher cost is justified if both strength and ductility are important. See Fig. 9-17.

Fig. 9-18 summarizes the relationship between

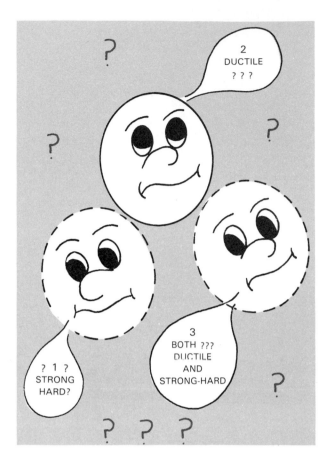

Fig. 9-17. If the steel user must have both strength and ductility, a more expensive steel is needed.

FAST COOLING OR QUENCHING	SLOW COOLING
Hardest	Softer
Strongest	Less Strong
Brittle	Ductile
Martensite	Ferrite-Pearlite Cementite
Small Grain Size	Large Grain Size

Fig. 9-18. Chart shows how cooling rate affects hardness, strength, ductility, steel structure, and grain size.

fast cooling and slow cooling, and the resulting hardness, strength, brittleness, ductility, and grain size.

AFFECTORS OF HARDNESS

Hardness basically results from two factors:
1. Speed of quench. The faster a steel is quenched, generally, the harder it will be after quenching.
2. Percent carbon. The more carbon the steel contains, generally, the harder the material will be after quenching.

Thus, if an extremely hard steel is desired, a high carbon content should be selected and a fast quenching method should be used in cooling the steel.

TEST YOUR KNOWLEDGE

Write your answers on a separate sheet of paper. Do not write in this book.

For the following nine questions, choose the best answer from the choices given.
1. Give the name of the structure of a solid solution of steel at 1200°F which contains a very small percentage of carbon (perhaps 0.02 percent). Assume that this steel has never been heat treated previously.
   a. Austenite   c. Ferrite
   b. Cementite   d. Pearlite
2. Name the structure of iron at room temperature which contains 6.67 percent carbon. Assume that this metal has never been heat treated.
   a. Ferrite   c. Pearlite
   b. Martensite   d. Cementite
3. Name the structure of a solid solution of steel at room temperature that contains 0.8 percent carbon. Assume that this steel has never been heat treated previously.
   a. Martensite   c. Austenite
   b. Pearlite   d. Cementite
4. What do you call a solid solution of steel at room temperature which has previously been rapidly quenched after having been heated above the upper transformation temperature, and containing a small percentage of carbon (perhaps 0.1 percent)?

a. Austenite   c. Martensite
b. Ferrite   d. Pearlite

5. Give the name of the structure of a solid solution of steel at room temperature which has previously been rapidly quenched after having been heated above the upper transformation temperature, and containing 0.8 percent carbon.
   a. Austenite   c. Pearlite
   b. Ferrite   d. Martensite

6. What is the hardest of the following types of iron-carbon?
   a. Martensite   c. Cementite
   b. Ferrite   d. Austenite

7. What is the most brittle of the following types of iron-carbon?
   a. Austenite   c. Ferrite
   b. Cementite   d. Martensite

8. Iron at 1800°F is rapidly quenched in water. What structural form of iron-carbon is the most predominant in the result if the alloy contains 0.9 percent carbon?
   a. Austenite   c. Martensite
   b. Cementite   d. Pearlite

9. Any steel with less than 0.8 percent carbon is called what?
   a. Eutectoid steel
   b. Hypoeutectoid steel
   c. Hypereutectoid steel

10. As grain size increases, which of the following increase, and which decrease?
    a. Hardness   c. Ductility
    b. Strength   d. Brittleness

11. The point on the iron-carbon phase diagram that marks the intersection of the upper transformation temperature line, the lower transformation temperature line, and the pearlitic carbon line is known as the _____ point.

12. Hardness is basically affected by two items.
    a. Speed of the _____.
    b. Percent _____.

13. There are eight letters on each of the iron-carbon phase diagrams shown in Fig. 9-19. These 16 letter locations correspond to 16 iron-carbon structures. Match each of these letters in the left column to one of the 22 iron-carbon structures listed in the right column. Some of these structures may be used more than once, some may not be used at all. Assume that all letters on the diagram involve slow heating and cooling only (no rapid cooling or quenching).

A. _____
B. _____
C. _____
D. _____
E. _____
F. _____
G. _____
H. _____
J. _____
K. _____
L. _____
M. _____
N. _____
P. _____
Q. _____
R. _____

1. All austenite.
2. All pearlite.
3. All martensite.
4. Almost all ferrite.
5. The maximum cementite possible in steel.
6. Half cementite and half austenite.
7. Half ferrite and half austenite.
8. Half pearlite and half cementite.
9. Half pearlite and half austenite.
10. Half ferrite and half cementite.
11. Half pearlite and half ferrite.
12. Over 80 percent austenite with some cementite.
13. Over 80 percent austenite with some ferrite.
14. Over 80 percent austenite with some pearlite.
15. Over 80 percent cementite with some austenite.
16. Over 80 percent cementite with some ferrite.
17. Over 80 percent ferrite with some austenite.
18. Over 80 percent pearlite with some austenite.
19. Over 90 percent pearlite with some cementite.
20. Over 80 percent pearlite with some ferrite.
21. Over 80 percent ferrite with some pearlite and some austenite.
22. A combination of austenite, pearlite, and cementite.

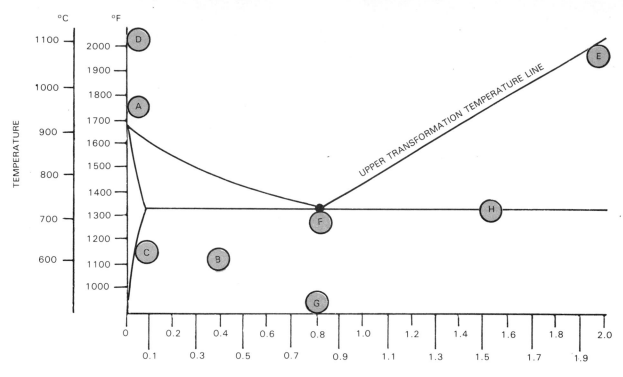

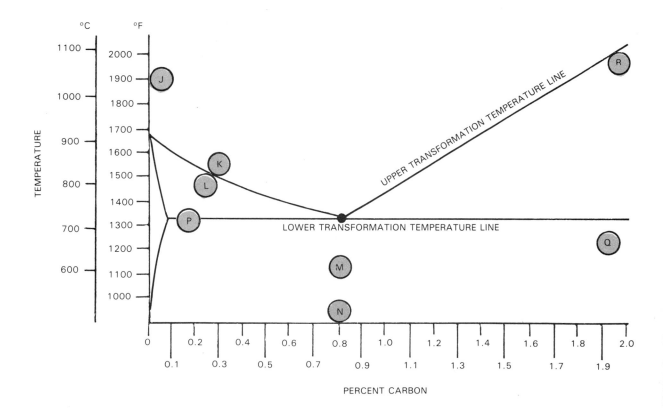

Fig. 9-19. Using listings given in Question 13, match each of the letter designations at left to one of the iron-carbon structures at right.

After studying this chapter, you will be able to:

☐ Compare the different structures of steel under a microscope.
☐ Discover how different ferrite, pearlite, cementite, austenite, and martensite look under the microscope.
☐ Recognize ferrite, pearlite, cementite, austenite, and martensite by looking at their microscopic pictures.
☐ Prepare a sample of metal so that it can be observed under the microscope.

LOOKING IN THE MICROSCOPE

When steel is magnified, it looks different. This can be seen by looking at it through the eye of a microscope, Fig. 10-1. Fig. 10-2 shows a view of a steel magnified 500X as observed in a microscope. The 500X means that the magnified view is 500 times larger than the steel itself.

EXAMPLES OF MICROSCOPIC APPEARANCES

Some examples of microscopic pictures of structures of steel are shown in Figs. 10-3 through 10-7. Note that the ferrite, pearlite, cementite, martensite, and austenite look very different from each other.

There are nicknames for each of these structures. The ferrite structure is called "patches." Pearlite looks like "ridges." Cementite resembles

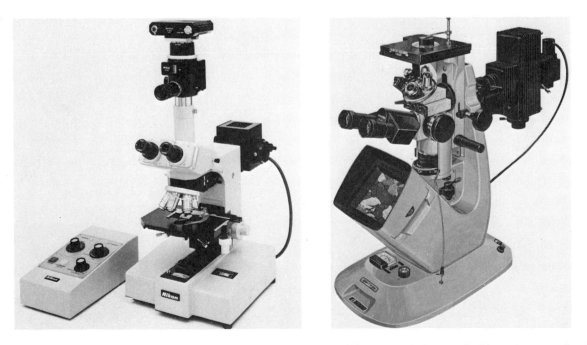

Fig. 10-1. Any of the metallurgical structures shown later in this chapter could be seen and photographed from these metallurgical microscopes. A — Nikon Metaphot Brightfield microscope. (Nikon, Inc., Instrument Div.)
B — Unitron Versament Metallograph. (Buehler Ltd.)

"white country roads." Martensite gives the impression of "needles." Austenite looks like "broken concrete slabs."

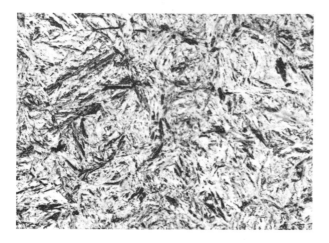

Fig. 10-2. This is a microscopic picture of 1095 steel martensite-bainite structure. Magnification is 500X. (Buehler Ltd.)

APPEARANCE OF FERRITE

Ferrite in the microscope appears white, Figs. 10-3 and 10-8. The small dark portions are the carbon or pearlite structures. If only ferrite appeared in a picture, it would be solid white.

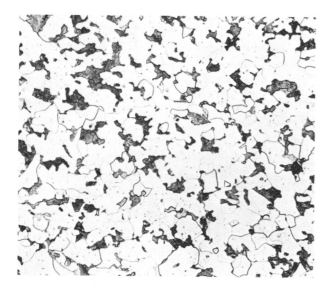

Fig. 10-3. This is 1020 steel ferrite structure with some pearlite. Magnification is 100X. (Buehler Ltd.)

APPEARANCE OF PEARLITE

In Chapter 9, you learned that pearlite is made up of ferrite and cementite. At 0.8 percent carbon, the ferrite and cementite content is balanced, and the entire metallurgical structure appears as pearlitic "ridges." The dark lines are cementite. The light colored ridges are ferrite. See Figs. 10-4 and 10-9.

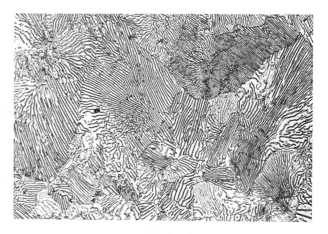

Fig. 10-4. This is 1095 steel pearlite structure. Magnification is 500X. (Buehler Ltd.)

APPEARANCE OF FERRITE-PEARLITE STRUCTURE

A composition made up of ferrite and pearlite is very interesting. The ferrite appears white. The pearlite appears dark or laminated. See Fig. 10-10.

When the carbon percentage is very small, there is much ferrite and little pearlite. Therefore, the appearance of the structure in Fig. 10-8 is very white.

As the carbon percent gets a little higher, and gets a little closer to pearlite, the microscopic picture appears darker, Fig. 10-11. You can see the pearlitic ridges in the dark portion of the picture.

A structure that is about half ferrite and half pearlite at 0.4 percent carbon shows about half white patches and half pearlitic ridges. See Figs. 10-12 and 10-13.

Compare these structures with pure pearlite, as shown in Fig. 10-14. Note how the white ferrite regions decrease as the carbon content increases. Note how the dark pearlitic regions increase as the carbon content increases. Note, too, how the pearlitic ridges become clearer as the dark regions become larger.

APPEARANCE OF
CEMENTITE-PEARLITE STRUCTURE

Cementite is the portion of Fig. 10-5 that appears like small, "white country roads." The pearlite still appears as its characteristic ridge pattern. In Fig. 10-4, where the carbon content is only slightly over 0.8 percent, there is almost pure pearlite. As the carbon content increases, in Fig. 10-5, "country roads" occupy a larger percentage of the space.

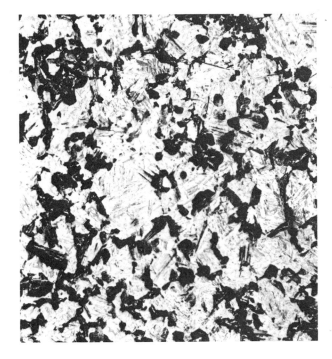

Fig. 10-6. This is 1045 steel martensite structure with some bainite. Magnification is 500X. (Buehler Ltd.)

Fig. 10-5. Cementite-pearlite structure. Magnification is 700X. (Struers, Inc.)

APPEARANCE OF MARTENSITE

Martensite takes many different microscopic appearances. However, all of them have a fine needlelike appearance. You get the impression of pointed lines whenever you look at a microscopic view of martensite. Those that look like small needles in Fig. 10-6 or 10-15 would look similar to the larger needles in Fig. 10-16 if the magnification of the microscope's lens were increased.

APPEARANCE OF AUSTENITE

At first, you may wonder how austenite could be photographed under a microscope, since it occurs only at elevated temperatures. It does not seem feasible to operate a microscope at 1700°F (927°C).

By adding special alloys to steel, austenite can be retained at room temperature. This is costly, of course, and is only done in special alloy steels. In regular low carbon steels or medium carbon steels, there would be no way to photograph austenite.

A microscopic view of stainless austenitic steel is shown in Fig. 10-7. It looks like broken slabs of an old concrete highway. These "slabs" slightly resemble ferrite. However, the ferrite formation has a rounder and more continuous curvature then austenite. The austenitic lines appear straighter and more abrupt. However, even though ferrite and austenite look a little bit alike in the microscope, their physical behavior is still far different.

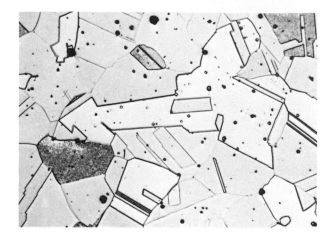

Fig. 10-7. This is 303 stainless steel austenitic structure. Magnification is 100X. (Buehler Ltd.)

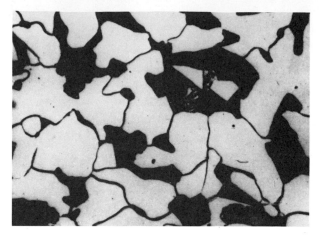

Fig. 10-10. Ferrite structure with some pearlite. Magnification is 700X. (Struers, Inc.)

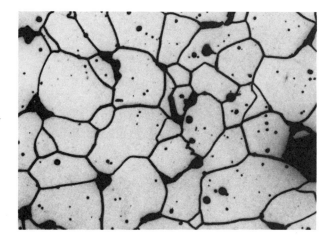

Fig. 10-8. Ferrite structure with some pearlite. Magnification is 700X. (Struers, Inc.)

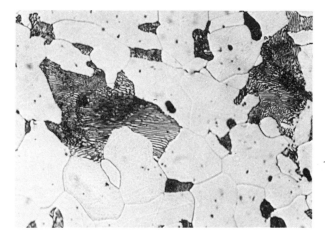

Fig. 10-11. This is 1018 steel ferrite structure with some pearlite. Magnification is 500X. (Buehler Ltd.)

Fig. 10-9. This is 1095 steel pearlite structure with some cementite. Magnification is 500X. (Buehler Ltd.)

Fig. 10-12. Ferrite-pearlite structure. Magnification is 700X. (Struers, Inc.)

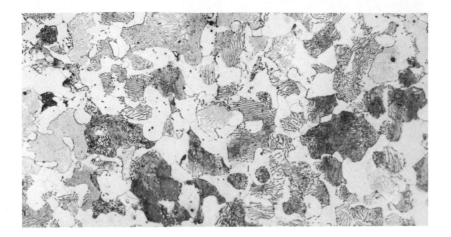

Fig. 10-13. This is 1045 steel ferrite-pearlite structure.
Magnification is 500X. (Buehler Ltd.)

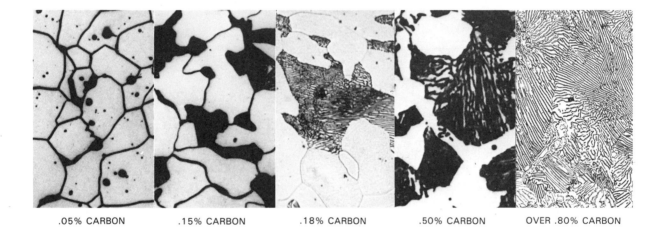

.05% CARBON   .15% CARBON   .18% CARBON   .50% CARBON   OVER .80% CARBON

Fig. 10-14. Five ferrite-pearlite structures with varying degrees of carbon. As carbon content decreases, more ferrite is present.
As carbon content increases, more pearlite is present.

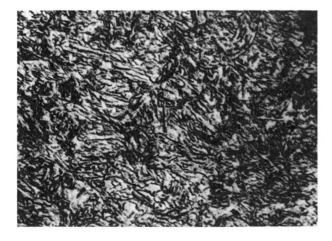

Fig. 10-15. Martensite structure. Magnification is 700X.
(Struers, Inc.)

Fig. 10-16. Martensite structure gives impression of "needles."
(Struers, Inc.)

SAMPLE PREPARATION PROCEDURE

The metal that appears in these microscopic photographs must be prepared and polished carefully before a good microscopic picture can be seen. The surface that is seen in the microscope must first be totally flat and smooth. Any irregularity will appear as a dark surface and will confuse the observer who attempts to analyze a structure under a microscope. In order to obtain this smooth flat surface, several preparatory steps are required. These include: grinding, molding, polishing, and etching.

1. Grinding — First the surface is ground to remove the rough scale and gross imperfections on the surface of the sample. See Fig. 10-17. Following rough grinding, fine grinding improves the surface until it begins to shine and reflect light slightly, Fig. 10-18.
2. Molding — The metal specimen is usually molded in plastic after rough grinding is completed. See Figs. 10-19 and 10-20. This makes the sample easier to hold throughout the polishing procedure. Generally, it is best not to mold before grinding. The sample can

overheat during grinding if your fingers are not touching it.

Fig. 10-18. Fine grinding further improves surface of sample. (Buehler Ltd.)

Fig. 10-19. A model Simplimet II hydraulic press is used to mount a metallurgical sample. (Buehler Ltd.)

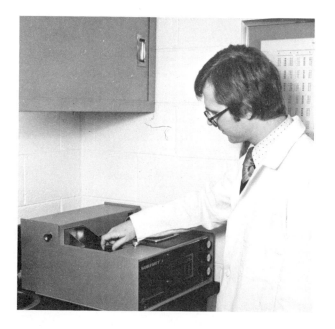

Fig. 10-17. Rough grinding is first step in preparing metal sample for microscopic examination. (Buehler Ltd.)

3. Polishing — Two types of polishing are normally involved. See Fig. 10-21. First, rough polishing, A, removes the imperfections that grinding has left. Then, fine polishing, B, will leave a mirrorlike finish on the surface of the steel with all scratches removed, Fig. 10-22.
4. Following grinding and polishing, the smooth polished surface is etched.

ETCHING

When acid is applied to the smooth metal surface, some metallic structures will be eaten away by the etching process more rapidly than others. The areas that are dissolved most rapidly will appear as dark shadows under the microscope. Those surfaces that react slowly to the acid appear light.

A

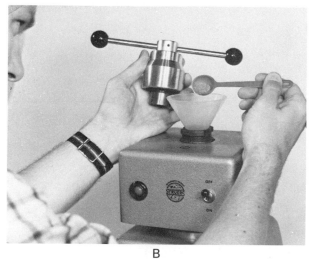

B

Fig. 10-20. A Prestopress is another type of hydraulic press used to mount a metallurgical sample. A — Lifting off upper assembly reveals molding area of press. B — Adding material is one of first steps for molding sample in plastic. (Struers, Inc.)

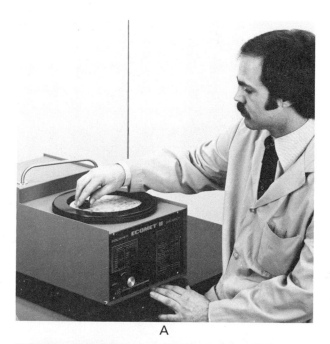

A

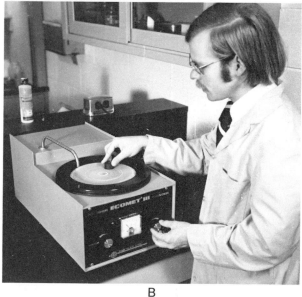

B

Fig. 10-21. Polishing of metallurgical sample follows molding operation. A — Rough polishing to remove imperfections grinding has left. B — Fine polishing to remove all scratches. (Buehler Ltd.)

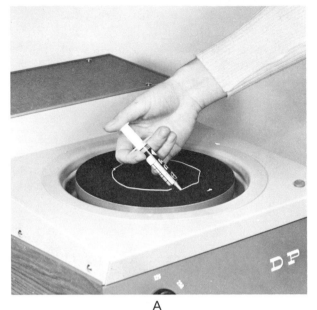

A

B

Fig. 10-22. Polishing equipment is shown: A—Polishing wheel. B—Polishing with an automatic specimen holder. (Struers, Inc.)

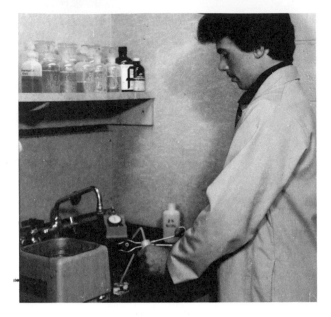

Fig. 10-23. First step in etching is alcohol rinse. (Buehler Ltd.)

2. You can apply the acid in many ways: by dripping acid onto the sample; by submerging the sample in a dish of acid; by swabbing an acid solution onto the sample. See Figs. 10-24 and 10-25. The time that the acid should be kept

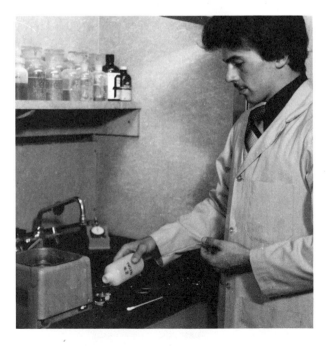

Fig. 10-24. Preparing acid before etching operation. (Buehler Ltd.)

The etching procedure involves the following steps:

1. Clean the surface with alcohol, Fig. 10-23, then let the sample dry in air. Alcohol evaporates rapidly. Wiping it off may leave smear marks.

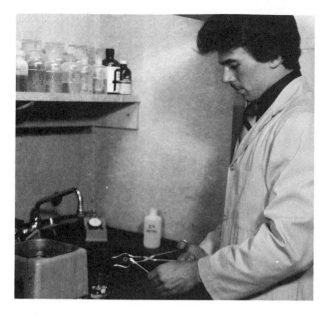

Fig. 10-25. Applying acid to sample during etching process. (Buehler Ltd.)

in contact with the sample varies considerably with every material. The time required may vary from a few seconds to a few minutes.

3. After the acid has had sufficient time to erode the metal surfaces, wash away the acid with water. It is best to put the sample under a stream of running water to remove all traces of acid.

4. As soon as you remove the sample from the water stream, wash it again with alcohol. This will help to prevent water marks. A 2 percent nitric acid solution in methyl or ethyl alcohol is often used. Good etching technique requires much practice.

After etching is completed, the sample is ready to be viewed under a microscope. Then, you can identify the microscopic structure of the steel. Fig. 10-26 compares the surface condition of a steel as it progresses through grinding, polishing, and etching.

LIGHT AND DARK SECTIONS

There is no absolute rule to use in determining which microscopic elements in the steel appear white and which elements appear dark.

This varies. However, there are a few general rules that usually apply.

The structure that has the greater percentage of iron or ferrite will generally appear white. The structure that contains the greater percentage of carbon will generally appear dark. There are exceptions to this rule.

Thus, the ferrite in the ferrite-pearlite structure of Fig. 10-10 or Fig. 10-12 appears white. The pearlite, which has the greater carbon content, appears dark.

In Fig. 10-4, the cementite portion of the pearlite appears darker. It has the higher percentage of carbon. The ferrite portion of the pearlitic structure appears white. It has less carbon.

One of the exceptions to the white-dark guideline is in the cementite-pearlite structure of Fig. 10-5. Here the cementite, which has the higher percentage of carbon, appears white and the pearlitic structure appears darker.

COMPARING THE MICROSCOPIC STRUCTURES OF DIFFERENT STEELS

It is interesting to compare similar structures of different types of steels.

In comparing 1018 steel to 1045 steel: 1045 steel has a higher percentage of pearlite because it contains approximately 0.45 percent carbon; 1018 steel contains approximately 0.18 percent carbon and therefore generates a lighter microscopic picture. Remember, as the carbon content increases, the amount of pearlite increases and the microscopic picture becomes darker.

In comparing 1018 steel and 1045 steel to other steels: 1018 steel would appear similar to the steel shown in Fig. 10-10 or Fig. 10-11; 1045 steel appears similar to the steel shown in Fig. 10-12 or Fig. 10-13.

In Figs. 10-27 through 10-31, you can see how the "white" ferrite structure gradually changes to

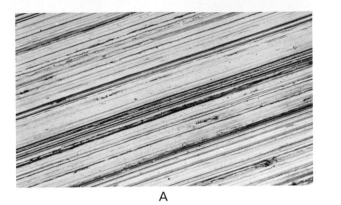

A

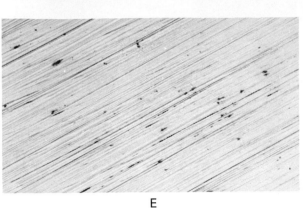

E

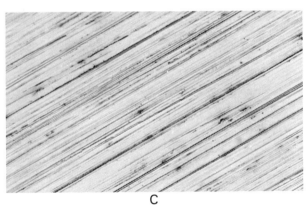

B

F

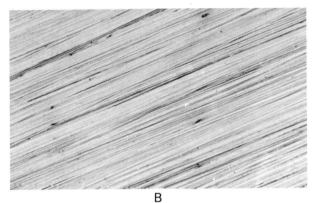

C

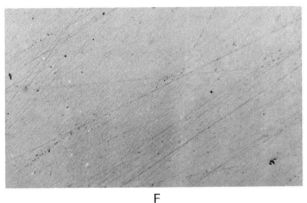

G

D

H

Fig. 10-26. Metallurgical surface is shown during eight stages of sample preparation. Magnification is 250X. A—After grinding with 180 grit paper. B—After grinding with 240 grit paper. C—After grinding with 320 grit paper. D—After grinding with 400 grit paper. E—After grinding with 600 grit paper. F—After rough polishing with 6 micron diamond abrasive on nylon cloth. G—After final polishing with .05 micron aluminum oxide on microcloth. H—After etching with .2 percent nital acid. (Buehler Ltd.)

"darker, ridge-like" pearlite as the carbon content increases. In Fig. 10-27, with only .06 percent carbon, the steel contains almost all ferrite. As the carbon content increases to 0.20 percent in Fig. 10-28, small colonies of pearlite ridges start to appear even though the ferrite structure is still predominant.

Since 0.36 percent carbon is almost exactly halfway between pure ferrite (0 percent carbon) and pure pearlite (0.80 percent carbon), Fig. 10-29 shows a structure that is approximately half ridges and half white ferrite. By the time we reach 0.53 percent carbon in Fig. 10-30, the pearlite has become predominant and the white ferrite is only in thin sections. In Fig. 10-31, at 0.86 percent carbon, the ferrite has disappeared.

The same five structures appear in Figs. 10-32 through 10-36, but at a lower magnification level. Comparing these five photographs from 0.06 percent carbon to 0.86 percent carbon, you again get the feeling of the white ferrite structure gradually disappearing as the carbon content increases and the pearlite ridges move in.

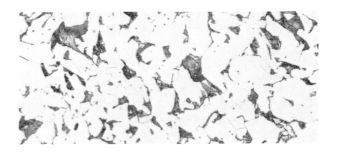

Fig. 10-28. Ferrite-pearlite structure with 0.20 percent carbon. Magnification is 1000X.

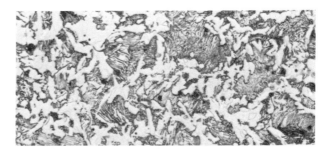

Fig. 10-29. Ferrite-pearlite structure with 0.36 percent carbon. Magnification is 1000X.

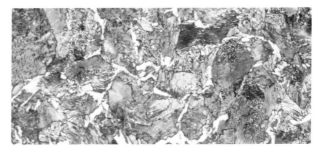

Fig. 10-30. Ferrite-pearlite structure with 0.53 percent carbon. Magnification is 1000X.

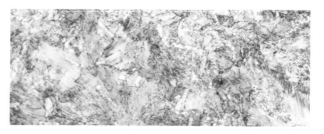

Fig. 10-31. Pearlite structure with 0.86 percent carbon. Magnification is 1000X.

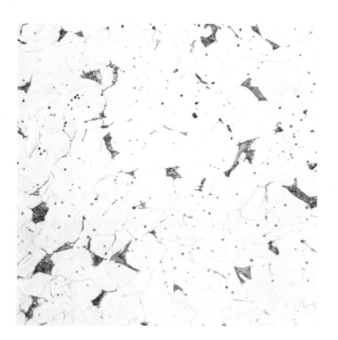

Fig. 10-27. Ferrite-pearlite structure with 0.06 percent carbon. Magnification is 1000X.
(Figs. 10-27 through 10-36 courtesy of LECO Corporation)

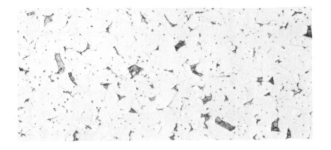

Fig. 10-32. Ferrite-pearlite structure with 0.06 percent carbon. Magnification is 500X.

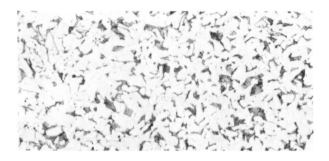

Fig. 10-33. Ferrite-pearlite structure with 0.20 percent carbon. Magnification is 500X.

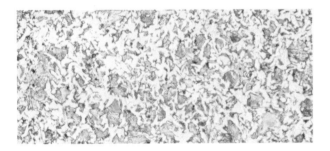

Fig. 10-34. Ferrite-pearlite structure with 0.36 percent carbon. Magnification is 500X.

Fig. 10-35. Ferrite-pearlite structure with 0.53 percent carbon. Magnification is 500X.

Fig. 10-36. Pearlite structure with 0.86 percent carbon. Magnification is 500X.

COMBINATION OF STRUCTURES

Fig. 10-37 shows a combination of steel structures:
1. The white areas are ferrite.
2. The lighter gray areas are martensite.
3. The laminated portions are pearlite.
4. The very darkest sections are bainite, a structure that will be discussed in Chapter 13.

Fig. 10-37. This is a microscopic picture of 1045 steel martensite-bainite structure with some ferrite and pearlite. Magnification is 500X. (Buehler Ltd.)

Microscopic Structures

TEST YOUR KNOWLEDGE

Write your answers on a separate sheet of paper. Do not write in this book.

1. What does 400X mean?
2. Which would show more detail and a closer view of a piece of metal, 50X or 200X?
3. Which steel structure under the microscope looks like "needles?"
4. Which steel structure under the microscope looks like "patches?"
5. A steel structure that is basically made up of "ridges" with a few small "white country roads" would be a combination of _____ and _____.
6. A ferrite-pearlite structure that appears almost entirely white might have:
   a. 0.1 percent carbon.
   b. 0.4 percent carbon.
   c. 0.7 percent carbon.
   d. 1.4 percent carbon.
7. A ferrite-pearlite structure that appears half white and half dark might have:
   a. 0.1 percent carbon.
   b. 0.4 percent carbon.
   c. 0.7 percent carbon.
   d. 2.0 percent carbon.
8. In a ferrite structure, the ferrite normally appears:
   a. White.
   b. Dark.
   c. Either.
9. When a metal sample is prepared for viewing under the microscope, several steps are required. List the following five steps in the correct chronological order:
   a. Etching.
   b. Fine grinding.
   c. Fine polishing.
   d. Rough grinding.
   e. Rough polishing.
10. As a general rule, a light section will indicate _____ (more/less) carbon.

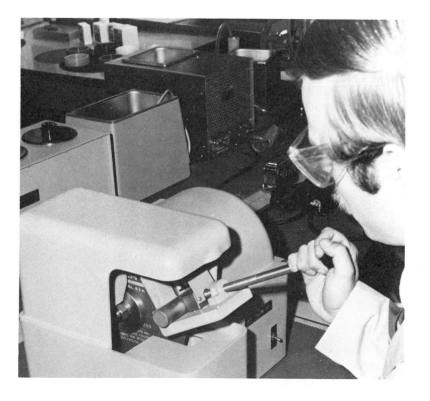

Samples to be heat treated are cut to size by a rotating abrasive cutting wheel.
(Buehler Ltd.)

11 HEAT TREATING AND QUENCHING

After studying this chapter, you will be able to:

☐ Explain what heat treating is.
☐ Tell what quenching is.
☐ Discuss the four stages that metal goes through as it is quenched.
☐ Name some different quenching liquids and how these different liquids affect the metal.
☐ Point out the good and bad effects of quenching a material more rapidly.
☐ Apply some practical techniques in quenching.

HEAT TREATING

HEAT TREATING can be classified as any metallurgical process that involves heating or cooling. It is a term widely used in industry. Heat treating includes many processes such as heating, quenching, annealing, normalizing, tempering, surface hardening, and any other heating or cooling process.

When a metal goes through several heating and cooling processes, the entire "recipe" is often referred to as the HEAT TREAT for that material.

Heat treating furnaces are of many different types, styles, and sizes. See Figs. 11-1 through 11-4.

QUENCHING

QUENCHING, by definition, is a controlled cooling process which causes metals to harden. Before quenching is done, the material must be heated to a high temperature. Quenching can be done from any elevated temperature. However, if hardness is important, the material should be heated to a temperature above the transformation region.

In quenching, parts may be inserted and removed from the oven or furnace individually by means of tongs, Fig. 11-5, or the parts may be carried in groups and lowered into the quenching tank. See Fig. 11-6.

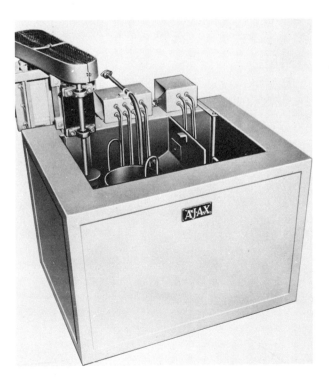

Fig. 11-1. This quenching furnace employs molten salt at temperatures of 350ºF to 750ºF. It is heated internally by electric resistance heating elements.
(Ajax Electric Company)

Fig. 11-2. This heat-treating furnace is electrically heated and is used for heat treating strip steel. (The Electric Furnace Company)

Fig. 11-3. Interior of this roller hearth heat-treating furnace has cast heating elements on top, bottom, and side walls. (The Electric Furnace Company)

Fig. 11-4. This heat-treating furnace is used for annealing coils of tubing, rod, strip, or wire and for annealing and heat treating motor and transformer laminations. (The Electric Furnace Company)

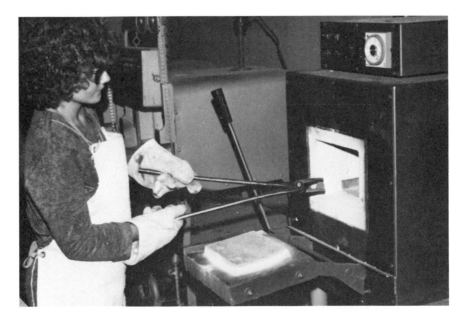

Fig. 11-5. Sample is removed from furnace by means of tongs.

The parts are plunged quickly into the quenching medium and submerged until they are cool. During the cooling operation, it is advisable to agitate or vibrate the parts very rapidly so cooling takes place as fast as possible.

There are several different quenching mediums. These include: water, brine, oil, air, molten salt, sand, and other less common ones. The most common quenching mediums will be discussed later in this chapter.

Fig. 11-6. Cylinders, held by a special fixture, are lowered into a quench tank. (J.W. Rex Company)

FOUR STAGES OF QUENCHING

Quenching action on a metal is not a simple process. See Fig. 11-7. There are four separate stages that the metal goes through as it cools from an elevated temperature down to room temperature during quenching:
1. Vapor formation stage.
2. Vapor covering stage.
3. Vapor discharge stage.
4. Slow cooling stage.

In the VAPOR FORMATION STAGE, the metal starts to cool but, soon, the cooling slows down as a vapor film starts to form. As soon as the metal is plunged into the liquid, the liquid next to the metal boils and causes a vapor film of bubbles to coat the outside surface of the metal. See Fig. 11-8.

During the VAPOR COVERING STAGE, this film of bubbles acts like a blanket or sleeping bag around the metal. The bubbles tend to stick to the metal and insulate it. Unless this vapor film is removed quickly, the metal will not cool rapidly enough.

To reduce the effect of this vapor blanket, the metal should be agitated as much as possible. This agitated shaking will cause some of the bubbles to fall off and cause the metal to begin to cool rapidly again. The longer the bubbles are permitted to stay on the metal, the slower cooling will take place and, therefore, less hardness and strength will result. Unless the film is removed completely and evenly, soft spots, warpage and cracking can also occur.

The VAPOR DISCHARGE STAGE is violent. It is at this point that the vapor film begins to collapse. As the vapor film collapses, it tends to explode off of the surface. This exploding or boiling action is violent enough to "rip off" the outer scale of the metal. This vapor discharge stage will

171

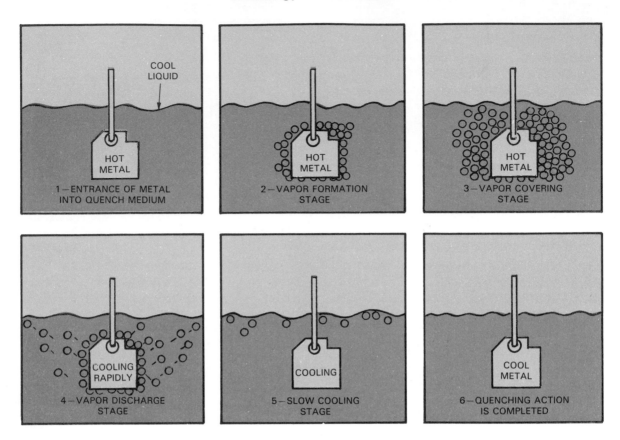

Fig. 11-7. Metal encounters four different quenching stages after its entrance into quench medium and before quenching action is completed.

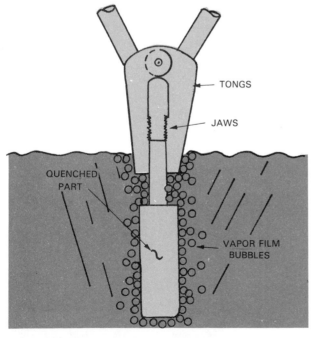

Fig. 11-8. When metal is plunged into a quenching medium, it is immediately covered with film of bubbles which cling to outside surface of metal.

generally occur quite quickly, sometimes in less than one second. The "fireworks" are loud enough to be heard by the ear. It can be recognized as a cracking or sizzling sound. In the old days, this sound was referred to as the "water biting the steel." The greatest amount of temperature drop takes place during this stage.

After the violence of the liquid discharge stage is over, the SLOW COOLING STAGE takes place. During this stage, the metal cools more slowly until it reaches room temperature. No further vapor film is formed. The cooling in this stage is much more gentle.

Fig. 11-9 plots the cooling of a typical material as it passes through all four cooling stages. Note from the curve that the greatest cooling takes place during Phase 3, the vapor discharge stage. During this stage, the temperature drops most drastically in a short time.

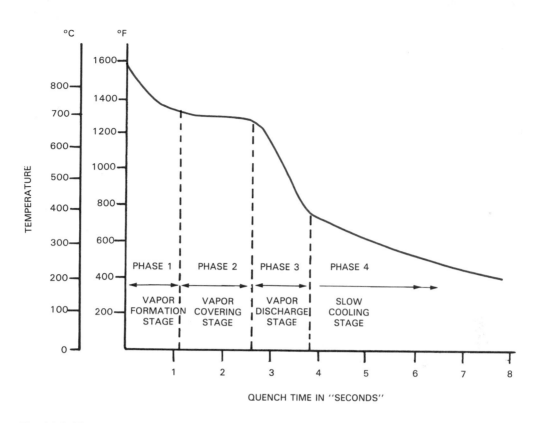

Fig. 11-9. Temperature drop and quench time in seconds are shown in four stages of quenching.

QUENCHING MEDIUMS

The term QUENCHING MEDIUM refers to the liquid into which the metal is plunged during the quenching process. Since all quenching mediums are not liquid, the term "medium" is used rather than "liquid." Some gases such as air are used in quenching. Some solids such as sand are occasionally used in quenching.

WATER QUENCHING

WATER is the most commonly used quenching medium. It is inexpensive, convenient to use, and provides very rapid cooling.

Water is especially used for low carbon steels, wherein a very rapid change of temperature is necessary in order to obtain good hardness and strength.

Water provides a very sudden, drastic quench. It can cause internal stresses, distortion, or cracking.

The water for quenching should be about room temperature for best results.

OIL QUENCHING

OIL is more gentle than water as a quench. Therefore, it is used for more critical parts, such as parts which have thin sections or sharp edges. Examples are razor blades, springs, and knife blades. Since oil is more gentle, there is less chance of internal stresses, distortion, or cracking. However, because it is more gentle, oil generally does not produce as hard or strong of a steel as water does. Therefore, the metallurgist must decide which is most important: (a) hardness and strength or (b) the elimination of cracking and distortion.

Oil gives a more effective quench when it is heated slightly above room temperature to 100°F or 150°F (38°C or 66°C). This may seem strange since normally the colder the quenching medium is, the harder and stronger the material will be when it is quenched. However, oil is very thick.

Heating it to 100°F will cause it to be thinner and permit it to circulate more easily as it quenches the part. Hence, warm oil will quench a part more rapidly than cold oil.

AIR QUENCHING

Air quenching is less drastic than either oil or water. The physical procedure involved in air quenching is illustrated in Fig. 11-10. The hot sample is placed on a screen, Fig. 11-11. Then, high-speed, cool air enters from below, passes through the screen, and strikes the hot metal parts. See Fig. 11-12.

Obviously, air does not cool parts as rapidly as either oil or water. This is both an advantage and a disadvantage. Because of the slower rate of cooling, there is less chance of internal stresses, distortion, and cracking. However, because the cooling rate is slower, the strength and hardness will not be as high unless special alloys are used in the metal.

Therefore, air quenching is generally used only on steels which have a very high alloy content. Special alloys such as chromium and molybdenum are selected because they are known to cause the material to harden even though a slower quenching method is used.

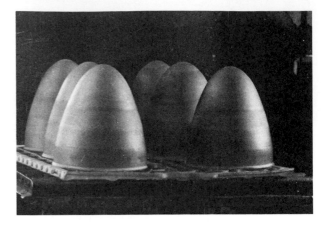

Fig. 11-11. High temperature alloy cones, just removed from furnace, are air quenched. (J.W. Rex Company)

BRINE QUENCHING

The effect of a BRINE QUENCH is very similar to the effect of a water quench. Brine will cool the material slightly faster than water, and the quenching action will be a little bit more drastic. The difference in the results of the two processes, however, is very slight.

Brine itself is salt water. A typical brine quenching medium contains 5 to 10 percent salt (sodium chloride) in water.

The reason for the slightly increased cooling rate is the action of the salt. The salt particles reduce the time of the vapor film stage. The salt precipitates out and causes the outer surface of the metal to explode off of the part. These tiny explosions cause disruption of the vapor layer. The vapor layer is removed more rapidly. With brine quench, more scale leaves the metal than with water quench.

Quenching in a molten salt bath is another quenching technique which should not be confused with brine quenching. In this method, parts are submerged in a molten salt bath which provides a rapid rate of heat transfer. For example, the furnaces in Fig. 11-1 employ molten salt at temperatures of 350°F to 750°F (177°C to 400°C).

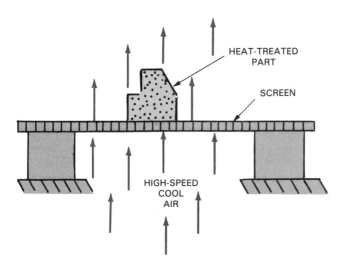

Fig. 11-10. In air quenching, high-speed cool air is rushed across hot part.

HEAT-TREATED PART

SCREEN

HIGH-SPEED COOL AIR

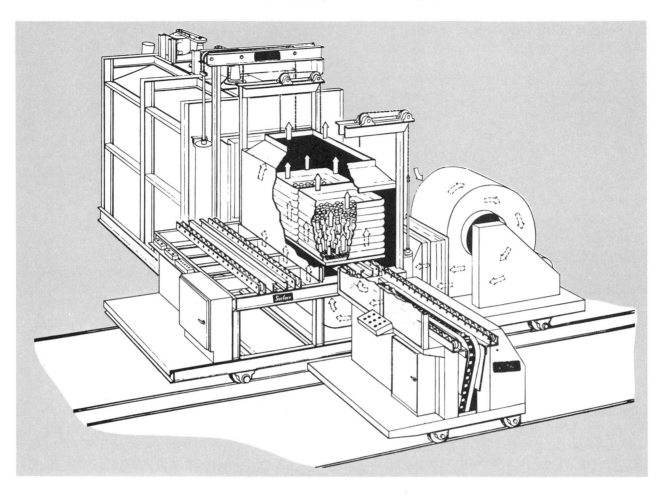

Fig. 11-12. Cutaway view reveals air quenching chamber. Loaded boxes or trays are pushed on conveyor into furnace at left rear of drawing. After heating and soaking at desired temperature, trays are moved on conveyor from furnace to air quenching chamber where parts are quenched by forced blast of air from blower. (Iron Castings Society)

SUMMARY OF EFFECTS OF QUENCHING MEDIUMS

Next, we will compare the advantages and disadvantages of the four quenching mediums we have discussed.

Speed of quench:
1. Water and brine are the fastest.
2. Oil is next fastest.
3. Air is the slowest.

Attainment of hardness and strength:
1. Water and brine produce greatest strength and hardness.

2. Oil produces next greatest strength and hardness.
3. Air produces least strength and hardness. However, as mentioned, the addition of expensive alloys in a metal can cause air-quenching steels to harden almost as much as those quenched in oil or water or brine.

Danger of internal stresses, distortion, and cracking:
1. Air is best because it is least drastic.
2. Oil is next best.
3. Water and brine are quenching mediums most likely to cause cracks or distortion or internal stresses.

Therefore, if you need to select the best quenching medium for a job, your first questions would be:

1. *Is it important that the part be hard and strong?* See Fig. 11-13.
2. *Would a certain amount of distortion or cracking be very detrimental to the part?* See Fig. 11-14. If hardness and strength are the most important, water or brine would be selected. If resistance to distortion and cracking are the most important, air would be used. Oil is a compromise or happy medium between the two. If both hardness and elimination of distortion are extremely important, you would use a more expensive alloy steel, and then air quench or oil quench the steel.

EFFECTS OF TEMPERATURE AND QUENCHING MEDIUM

Temperature of the water or oil has a definite effect on the effects of quenching. In Fig. 11-15, it can be seen that water at 70°F (21°C) can cool down the metal in almost half the time that 120°F (49°C) water requires. In many cases, this will be the difference in attaining a metal that has a satisfactory hardness or not.

Fig. 11-16 compares several different quenching mediums. Note that the fastest quench is from the brine or water. The slowest quench is from still air. Oil, as expected, is partway between.

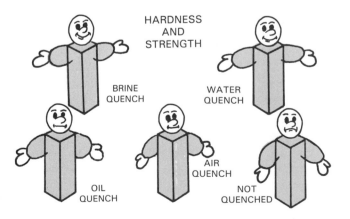

Fig. 11-13. Greatest hardness and strength for most materials are obtained by brine quenching or water quenching.

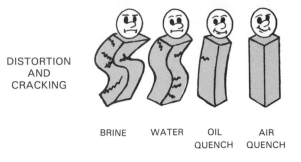

Fig. 11-14. Least distortion and cracking for most materials is obtained by air quenching.

PRACTICAL TECHNIQUES IN QUENCHING

A metallurgist must be a very practical individual because many practical aspects of quenching must be kept in mind when selecting a quenching process. There are many questions to consider before starting a quenching operation:

1. *How long should the material be left in the quenching tank?* The rule of thumb is to leave the sample in the quenching medium until it is cool enough to be touched by hand.
2. *How long must the part be left in the oven before removing it for quenching?* The rule of thumb is that after the oven is at the desired heating temperature, the part is left in the oven for one hour for every inch of thickness. Thus, if a metal cube were being heat treated and it measured 2 in. by 2 in. by 2 in., it would be left in the oven at least two hours. If a part measured 1/4 in. by 1/4 in. by 1/4 in., it need only be left in the oven for 15 minutes before quenching. See Fig. 11-17.
3. *How important is agitation?* Sometimes agitation is just not practical because there is a large quantity of parts or because a part is awkward to handle. However, the effects of agitation are quite apparent, as shown in Fig. 11-18.

With oil, agitation does not provide as much benefit as it does with water. Oil bubbles tend to cling more tenaciously to the metal surface. In a brine or water quench, a second or two of shaking will do a lot of good toward dislodging the vapor bubbles.

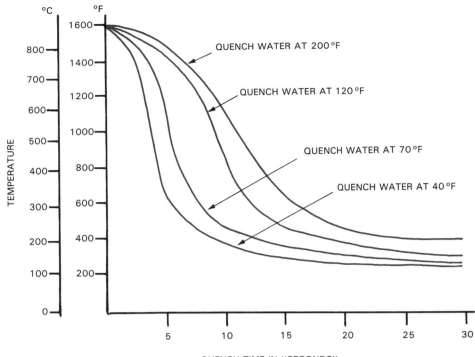

Fig. 11-15. Temperature of quenching medium has an effect on quench time.

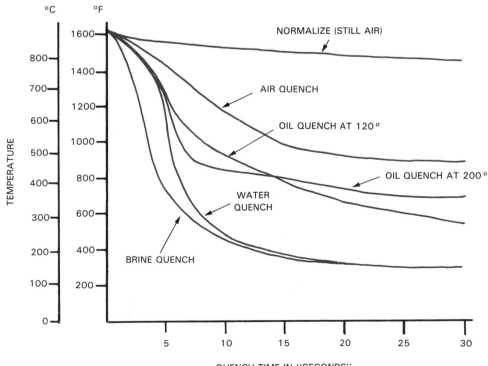

Fig. 11-16. Graph compares quenching rates with different quenching mediums.

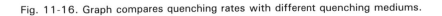

Since the vapor film phase lasts only a second or two, there is little benefit in agitating or shaking the part after it has been in the quenching medium more than a few seconds.

4. *How do I avoid distortion in thin parts?* When a part has a thin web between heavier sections, Fig. 11-19, there is a good chance of distortion during quenching. To reduce the chance

of this, there are special clays that can be put on the part during heating and quenching, See Fig. 11-20. These clays protect the part and reduce the severity of the quenching action when the metal first enters the liquid. This reduces the chance of cracking and distortion in thin projections and even in sharp corners like those shown in Fig. 11-21.

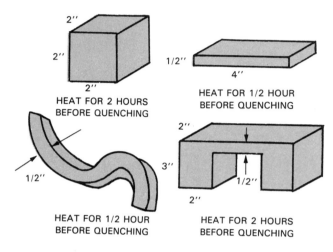

Fig. 11-17. A part should be left in oven at least one hour for every inch of thickness before quenching.

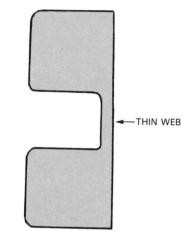

Fig. 11-19. Part with thin web between heavier sections is susceptible to distortion during quenching.

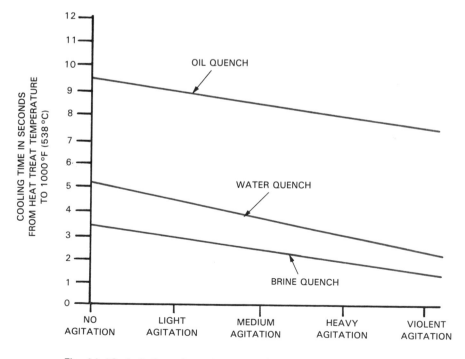

Fig. 11-18. Agitation of part being quenched reduces cooling time.

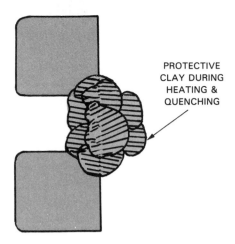

Fig. 11-20. Special clays can be put on thin webs during heating and quenching to reduce chance of distortion.

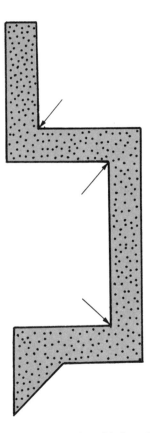

Fig. 11-21. Special clays can be added to sharp corners to reduce chance of cracking and distortion.

TEST YOUR KNOWLEDGE

Write your answers on a separate sheet of paper. Do not write in this book.

1. Which of the following processes could be classified as heat treating?
   a. Heating.
   b. Water quenching.
   c. Oil quenching.
   d. Full annealing.
   e. Process annealing.
   f. Normalizing.
   g. Tempering.
   h. All of these.

2. In addition to water quenching, name four other methods of quenching.

3. During the four stages of quenching, which stage appears to be the most violent?

4. During which of the four stages of quenching do bubbles form a blanket around the sample?

5. Which of the following four quenching methods is the most drastic?
   a. Air quench.
   b. Brine quench.
   c. Oil quench.
   d. Water quench.

6. Which of the following four quenching methods is the most commonly used?
   a. Air quench.
   b. Brine quench.
   c. Oil quench.
   d. Water quench.

7. Which of the following four quenching methods is the most gentle:
   a. Air quench.
   b. Brine quench.
   c. Oil quench.
   d. Water quench.

8. What is brine?

9. Which will provide the fastest quench, water at 75°F (24°C) or water at 120°F (49°C)?

10. What is the minimum amount of time that a sample should be left in an oven before quenching if the sample measures 3 in. x 3 in. x 3 in.?

11. What is the minimum amount of time that a sample should be left in an oven before quenching if the sample measures 1/2 in. x 1 in. x 2 in.?

12 ANNEALING AND NORMALIZING

After studying this chapter, you will be able to:

☐ Tell what annealing and normalizing are.
☐ Explain how annealing and normalizing affect hardness, strength, and brittleness.
☐ State the purpose of annealing and normalizing.
☐ Describe how annealing and normalizing affect the crystal structure of metal.
☐ Identify three different types of annealing.
☐ Point out the different ways in which annealing and normalizing affect metal compared to quenching.

WHAT ARE ANNEALING AND NORMALIZING?

In Chapter 11, you learned that quenching metal makes it harder and stronger. However, quenching also makes the material more brittle and less ductile. Sometimes it is more important that a material be ductile rather than hard and strong. This is where annealing and normalizing come in. They make steel softer, and more ductile. This causes the steel to be easier to machine, easier to bend and form, and less prone to cracking and distortion.

Fig. 12-1. Many tubes (foreground) are simultaneously annealed in this roller hearth furnace. (The Electric Furnace Company)

The definition of both ANNEALING and NORMALIZING would be "heating a metal above a critical temperature, and slowly cooling it to room temperature, in order to obtain a softer and less distorted material." Just as a fast quench produces a hard part, a slow cooling process produces a soft and more ductile part. Both annealing and normalizing involve slow cooling. Annealing furnaces are shown in Figs. 12-1 and 12-2.

NORMALIZING

The process of normalizing involves heating the material above the upper transformation temperature and then cooling it slowly *at room temperature*. In other words, when the material is removed from the furnace or oven, it is set out to cool slowly on a bench or a floor. Large parts may even be left to cool slowly outside. The steel is *not* submerged in a liquid. It is *not* quenched. The steel merely cools to room temperature with no interference from any external influence. After the material has cooled for a long time, the entire part will be at room temperature and the material is said to be "normalized."

ANNEALING

Annealing is very similar to normalizing except that cooling takes place still more slowly. In an-

Fig. 12-2. Metal strip is continuously annealed in this catenary furnace. (The Electric Furnace Company)

nealing, the steel is cooled in a temperature controlled oven. The metal first is heated above the upper transformation temperature or some other high temperature, and left to soak at that temperature for a period of time. Next, it is placed in an annealing oven where the temperature is slowly lowered. Since the temperature of this oven is lowered slowly, annealed steels cool at a slower rate than normalized steel.

The oven may be the same one in which the steel was heated, or the steel may be transferred to a second oven. When the oven temperature finally is lowered to room temperature and the material reaches room temperature, the part is said to have been "annealed." Annealing sometimes takes as long as several days in order to get a material as soft as possible.

EFFECTS OF ANNEALING AND NORMALIZING

Annealing and normalizing generally make a material less hard. Annealing and normalizing generally make a material less strong. They make a material less brittle and more ductile.

Annealing and normalizing reduce the amount of internal stresses in a material. This, in turn, reduces the tendency of the material to distort and crack.

Fig. 12-3 summarizes the given six characteristics of annealing and normalizing, and compares these processes to quenching.

PURPOSES FOR ANNEALING AND NORMALIZING

In the study of metallurgy, there is so much emphasis on attaining hardness and strength that one starts to consider hardness and strength as the "good guys" and softness and ductility as the "bad guys." This is not always true. There are many purposes in making a material softer and less strong:

1. So that the material is easier to machine.
2. To provide a material that is easier to form.
3. To relieve internal stresses.
4. To refine the crystal structure.

ANNEALING AND NORMALIZING	QUENCHING
Softens	Hardens
Weakens The Material	Strengthens
Causes Ductility	Causes Brittleness
Removes Internal Stress	Causes Internal Stress
Removes Distortion Trends	Causes Distortion
Removes Cracking Trends	Causes Cracking
Is A Slow Cooling Process	Is A Fast Cooling Process

Fig. 12-3. Purposes of annealing and normalizing are different than purposes of quenching.

MACHINEABILITY

A softer and more ductile material is easier to machine in the machine shop. Often, as much machining as possible will be done in the annealed or normalized condition even on a part that requires strength. Then, after most of the machining is completed, the part is hardened by heating and quenching. Only the last finishing cut is taken in the hardened condition. This saves considerable time and cost in the machining department and reduces tool wear.

FORMING

A part that has been annealed or normalized also will respond better in the forming department. Processes such as spinning, rolling, bending, and deep drawing require considerably less machine horsepower if the part has been softened. Also a softer, more ductile part will be much less likely to crack or distort if it has previously been annealed or normalized before forming.

INTERNAL STRESSES-DISTORTION-CRACKING

A material can build up internal stresses. Inside a piece of steel, stresses are created in the same way that internal mental stresses are created in a human being when things go wrong. If these internal stresses are allowed to remain in a metal, the part will eventually distort or crack. Just as a human being who feels tension can take a hot shower to relieve stress and relax internal nerves, metal too can be relaxed by annealing or normalizing in a hot oven.

Internal stresses in a metal may be caused by previous processing operations such as welding, cold working, casting, forging, punching, drawing, extruding, or machining.

REFINING THE CRYSTAL STRUCTURE

Another reason for annealing or normalizing is to change the crystal shape. After some types of metalworking, particularly cold working, the metal crystals are elongated or stretched. Annealing or normalizing can change the shape of these crystals back to normal when the material reaches the austenitic region. Then, upon cooling, a more desirable shaped crystal is redeveloped.

TYPES OF ANNEALING

There are three basic types of annealing:
1. Full annealing.
2. Process annealing.
3. Spheroidizing.

FULL ANNEALING

In full annealing, the steel is heated to a high temperature, Fig. 12-4. This temperature is normally $50° - 100°F$ ($10° - 38°C$) above the upper transformation temperature. Upon reaching this temperature, the steel changes to austenite and becomes face centered cubic. It is held at this temperature for a long time. A rule of thumb is to soak the material for one hour at the annealing temperature for every inch of its thickness. See Figs. 12-1 and 12-2.

The material then is slowly cooled in the oven. A cooling rate of $100°F$ per hour is typical for full annealing. A much slower cooling rate is used when time is not important, and when the elimination of internal stresses is especially critical.

It is best not to heat the material much above the upper transformation temperature or the grain size will become excessively large. The higher that steel is heated above the upper transformation temperature, the larger the grain size will be. A temperature of $50° - 100°F$ above the upper transformation temperature is normally sufficient to produce austenite without generating large grain size.

Full annealing might be referred to as "just plain annealing." If someone refers to a process as "annealing," chances are they are referring to "full annealing" rather than "process annealing" or "spheroidizing."

PROCESS ANNEALING

One major problem with full annealing is the time that it takes. In order to attain completely stress-free material, the part may need to be cooled slowly over an entire day or, occasionally, over several days. Not only does this tie up the parts but it also ties up ovens for long periods of time.

PROCESS ANNEALING is used when time is more important than full softening. In process annealing, the part is not softened or relieved as completely as in full annealing, but the time required is considerably less. Thus, process annealing is somewhat of a compromise between full annealing and not annealing at all.

During manufacture, internal stresses are built up in the parts. This is due to the processing

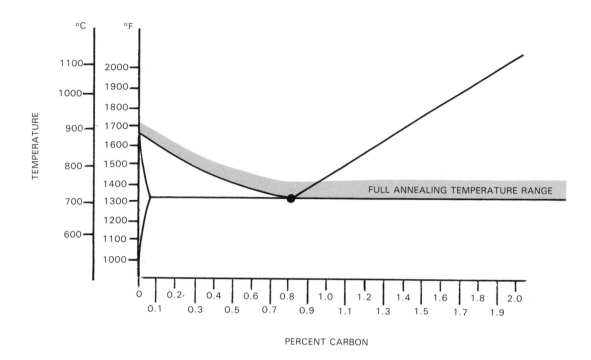

Fig. 12-4. In full annealing, steel is heated to high temperature.

forces discussed earlier (welding, cold working, etc.). It is not always necessary to get rid of 100 percent of these internal stresses if the time required to do that is excessive. If most of the stresses can be eliminated in a short period of time, process annealing is a practical economic solution.

In process annealing, the part is heated to only 1050° – 1300°F. Then it is cooled. This will STRESS RELIEVE the part. It will get rid of much of the internal stress, but not all of it. See Fig. 12-5.

Process annealing is frequently used as an intermediate heat-treating step during the manufacture of a part. A part that is stretched considerably during manufacture may be sent to the annealing oven three or four times before all of the stretching is completed. Thus: a part may be stretched a little, then process annealed to eliminate some internal stresses; then stretched

a little more, then process annealed again, etc.

After a rough machining cut, a part may be process annealed to remove most of the stresses caused by the cutting action. After stress relieving, a thin final cut may be taken which produces little additional stress. Thus, the complex part will end up in a nearly unstressed condition. See the bottom view in Fig. 12-6.

SPHEROIDIZING

SPHEROIDIZING is a quick annealing method almost identical to process annealing. Parts are heated to a temperature below or near the lower transformation temperature and then slowly cooled, Fig. 12-7.

The name "spheroidizing" comes from the microscopic appearance of the steel after it has been annealed. Many tiny spherical forms occur throughout the microscopic structure when

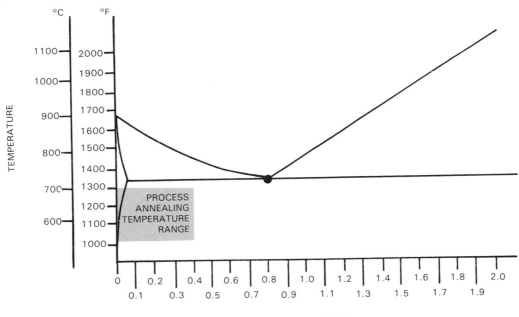

Fig. 12-5. In process annealing, steel is heated to temperature that is less than lower transformation temperature. This relieves some internal stresses without taking great deal of time.

cementite is process annealed, Fig. 12-8; hence, the name spheroidizing.

Thus, when cementite is process annealed and the globular forms appear in the microscopic pattern, it is referred to as "spheroidizing." When ferrite is annealed and globular forms do not appear, it is referred to as "process annealing." Spheroidizing involves high carbon steel.

NORMALIZING

In NORMALIZING, the steel is also heated above the upper transformation temperature, Fig. 12-9. It is then cooled in still air at room temperature. See Fig. 12-10.

The name "normalizing" comes from the original purpose of the process which was to return the steel to the "normal" condition that it was before cold working or other processing was done to it.

Normalizing does not leave the material as soft as full annealing does. It does not leave the material as ductile or internally stress-free as full annealing does. A normalized part will normally be a little stronger, harder, and more brittle than a part that has been full annealed. A normalized part will have a slightly finer, smaller grain structure.

COMPARISON OF ANNEALING, NORMALIZING, AND QUENCHING

The diagram in Fig. 12-11 compares annealing, normalizing, and quenching. At the far left of the chart is the extremely hard and strong region. The far right of the chart depicts the extremely soft and ductile region.

You will note that full annealing will render a piece of steel as soft as can be attained. A brine quench or water quench will render the material to the hardest possible condition.

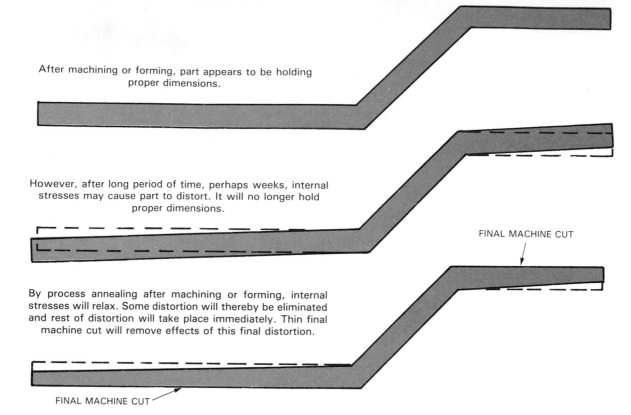

After machining or forming, part appears to be holding proper dimensions.

However, after long period of time, perhaps weeks, internal stresses may cause part to distort. It will no longer hold proper dimensions.

FINAL MACHINE CUT

By process annealing after machining or forming, internal stresses will relax. Some distortion will thereby be eliminated and rest of distortion will take place immediately. Thin final machine cut will remove effects of this final distortion.

FINAL MACHINE CUT

Fig. 12-6. Process annealing can eliminate hazard of unexpected distortion taking place after long period of time.

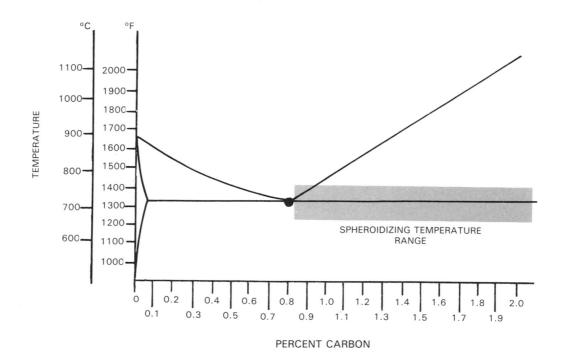

SPHEROIDIZING TEMPERATURE RANGE

PERCENT CARBON

Fig. 12-7. In spheroidizing, parts are heated to temperature near lower transformation temperature.

186

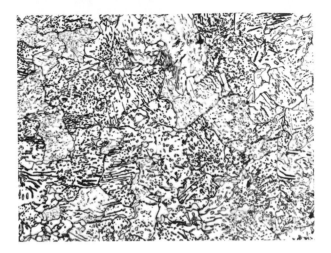

Fig. 12-8. Note spherical forms in 1095 steel spheroidized structure. (Buehler Ltd.)

Process annealing and spheroidizing cannot be compared on the overall scale since they are dependent on the condition of the metal before it is process annealed. Full annealing or normalizing can reduce the hardness of any steel to a very soft condition, even if it is a water-quenched part that measures harder than 60 R_c.

Fig. 12-9. These drill rods are being transferred to air-cooling station after being removed from this 22-ft. high, bottom-opening gantry furnace. (J.W. Rex Company)

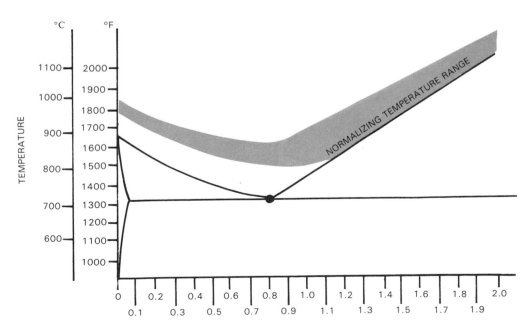

Fig. 12-10. In normalizing, steel is heated to high temperature where it changes to austenite. Then it is cooled slowly at room temperature.

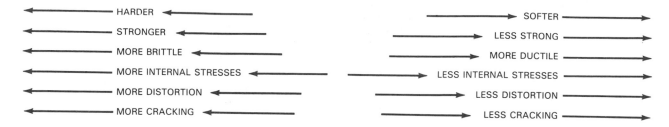

Fig. 12-11. When quenching, normalizing, and full annealing are compared, it is obvious that quenching produces relatively hard, strong, and brittle parts, while normalizing and full annealing produce softer, more ductile, and more stress-free parts.

TEST YOUR KNOWLEDGE

Write your answers on a separate sheet of paper. Do not write in this book.

1. How is a material cooled when normalizing?
2. What are the benefits of annealing or normalizing?
3. Which of the following will be true of steel that has been annealed or normalized?
   a. Stronger.
   b. More ductile.
   c. Harder.
   d. Less susceptible to cracking.
   e. More machinable.
   f. Easier to bend.
4. Which is faster?
   a. Full annealing.
   b. Process annealing.
   c. Normalizing.
5. In which process is the material heated the hottest?
   a. Full annealing.
   b. Process annealing.
   c. Spheroidizing.
6. "Just plain annealing" normally means?
   a. Full annealing.
   b. Process annealing.
   c. Spheroidizing.
7. What is the advantage of process annealing over full annealing?
8. Which takes longer to cool, a part that is being normalized, full annealed, or spheroidized?
9. After a part has been heated to a high temperature, which of the following three cooling processes would cause the part to become the softest?
   a. Full annealing.
   b. Normalizing.
   c. Quenching.
10. After a part has been heated to a high temperature, which of the following three cooling processes would cause the part to become the strongest?
    a. Full annealing.
    b. Normalizing.
    c. Quenching.

13 ISOTHERMAL TRANSFORMATION DIAGRAMS

After studying this chapter, you will be able to:

☐ Explain what an isothermal transformation diagram is.

☐ Cite the value of an isothermal transformation diagram and how it can be used.

☐ Tell if a material will become stronger by looking at its isothermal transformation diagram.

☐ Evaluate several industrial isothermal transformation diagrams.

☐ Recognize that different steels have very different isothermal transformation diagrams.

☐ Plot a time-line on an isothermal transformation diagram.

INTRODUCTION TO ISOTHERMAL TRANSFORMATION DIAGRAMS

The iron-carbon phase diagram is a very useful tool in metallurgy. When the percent carbon and the temperature are known, the structure of a particular steel can be established.

One element is missing from the iron-carbon phase diagram, however. That is *time*. If a steel is heated above the upper transformation temperature and cooled *slowly,* it becomes ferrite, or pearlite, or cementite. If a steel is heated above the upper transformation temperature and quenched *rapidly,* it becomes martensite.

The iron-carbon phase diagram cannot tell you what cooled *slowly* means or what quenched *rapidly* means in terms of specific time values. If steel is cooled from a high temperature to room temperature in six sec. is this rapid or is this slow?

If a material is cooled medium fast, what percent martensite occurs and what percent ferrite,

pearlite, or cementite occurs? The iron-carbon phase diagram cannot tell you this.

Thus, while the iron-carbon phase diagram is extremely valuable in some cases, it shows only a two-dimensional picture. It is unable to show us the effects of time.

BASICS OF I-T DIAGRAM

The ISOTHERMAL TRANSFORMATION (I-T) DIAGRAM *does* take time into account. It also is a very useful tool. It also has limitations.

The I-T DIAGRAM is a graph of temperature versus time for the process of cooling metal. From this diagram, the final structure of metal can be predicted.

I-T diagrams are also called "T-T-T diagrams" or "C curves" or "S curves." I-T stands for "isothermal transformation." T-T-T stands for "time-temperature-transformation." "C" or "S" refers to the shape of the curve.

Fig. 13-1 shows a skeleton of an I-T diagram. Note that temperature is plotted versus time. The two horizontal lines at the top of the diagram represent the UPPER TRANSFORMATION TEMPERATURE and the LOWER TRANSFORMATION TEMPERATURE. Therefore, to be transformed completely, steel must be heated above both of these lines.

The TIME SCALE at the bottom of the curve is a logarithmic scale. This means that the time increment is not constant along the scale. The time scale between one and 10 sec. is longer than the time between 1001 and 1010 sec. This type of scale is very handy to use when most of the action takes place early, and when changes at higher values of time take place rather slowly. In quenching, most of the "excitement" takes place between one and 10 sec. Therefore, the time-scale is "stretched out" between one and 10 sec.

LIMITATIONS OF I-T DIAGRAM

While the I-T diagram has the advantage over the iron-carbon phase diagram of getting time into the act, it also has a limitation. It does *not* plot percent carbon. Therefore, a new I-T diagram is necessary every time that the percentage of carbon changes. In fact, a different I-T diagram must be plotted for every type of steel.

TEMPERATURE-TIME-LINE

In Fig. 13-2, a TEMPERATURE-TIME-LINE

is added to the skeleton of the I-T diagram. This line follows the path of the temperature of the steel after quenching begins. In Fig. 13-2, the steel is heated to 1700°F (927°C) at point A and then quenched. When the steel reaches 700°F (370°C) at point B, it is held at that temperature. Quenching from point A to point B takes 10 sec.

The steel is held at 700°F for 90 more sec., point B to point C in Fig. 13-2. Finally, it is cooled suddenly to room temperature at point D.

In Fig. 13-3, time-line A shows a steel that is quenched rapidly. It reaches room temperature in only two sec. Temperature-time-line B shows a steel that is cooled very slowly. It takes 10,000 sec. or almost three hours before it reaches room temperature. Therefore, steel A would become a hard, strong steel, probably martensite. Steel B, which was slow cooled, becomes either ferrite, or pearlite, or cementite.

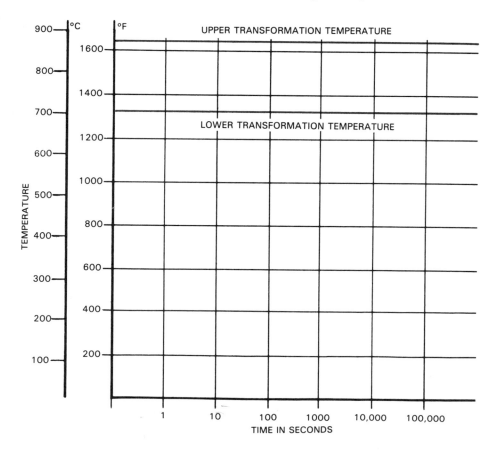

Fig. 13-1. Isothermal transformation diagrams plot temperature VS time.

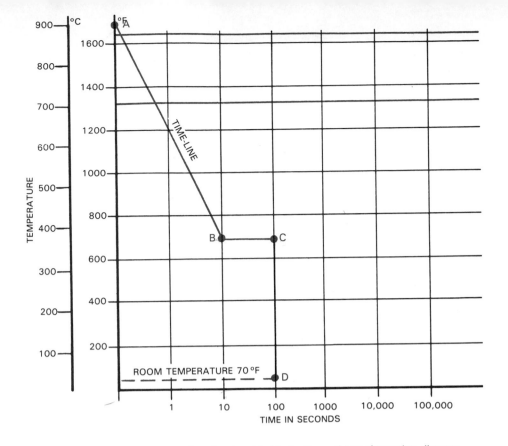

Fig. 13-2. A temperature-time-line is added to isothermal transformation diagram.

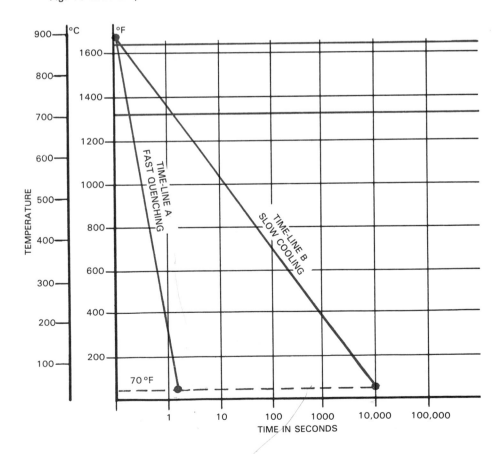

Fig. 13-3. On this isothermal transformation diagram: time-line A represents fast quenching; time-line B represents slow cooling.

C CURVES

Figs. 13-4, 13-5, 13-6, and 13-7 show the C CURVES for four different types of steel. You can use these C curves to determine whether a steel will be changed to martensite, ferrite, pearlite, or cementite, or some other structure. These C curves were originally developed by steel companies through repeated industrial tests and analyses of test results.

Publishing organizations have compiled the C curves in book form, and these books are commercially available.

PURPOSE OF I-T DIAGRAMS

Fig. 13-8 shows a simplified I-T diagram. It is essentially the same type of curve as the industrial I-T diagrams in Figs. 13-4, 13-5, 13-6, and 13-7, but shown in abbreviated form. The left C curve indicates the beginning of structural transformation. Thus, when the time-line of any steel is at the left of both C curves, the steel is 100 percent austenite and has not begun its transformation. After the time-line crosses over and is to the right of the C curves, it indicates that the steel has been completely transformed and no longer contains any austenite.

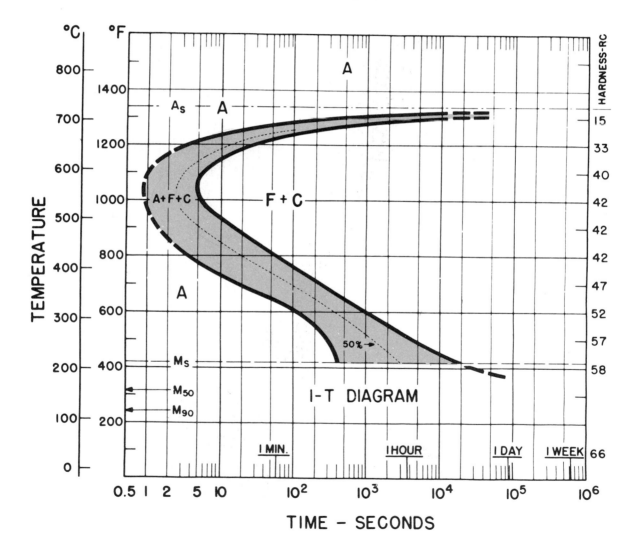

Fig. 13-4. This industrial I-T diagram is for 1095 steel.

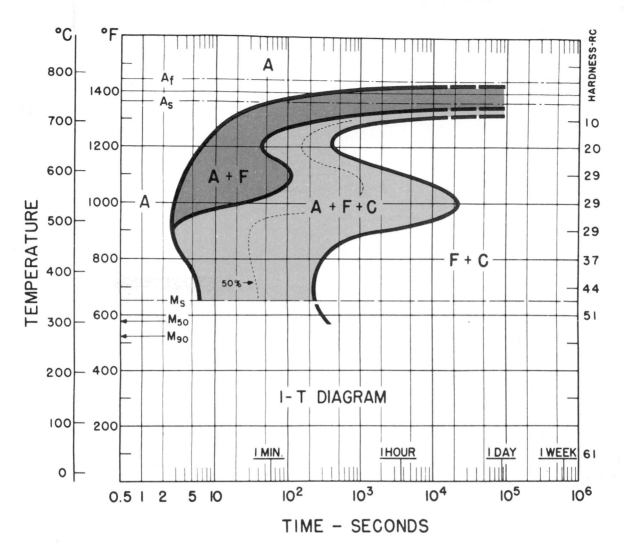

Fig. 13-5. This industrial I-T diagram is for 4140 steel.

The region between the left C curve and right C curve is the REGION OF TRANSFORMATION where the action takes place. It is here that austenite is converted to some other structure such as martensite or pearlite. The dotted line between the left and right C curve is the 50 percent transformation line. When a time-line for a steel reaches this line, 50 percent of the austenite has been transformed to something else and only 50 percent austenite remains. For example, point B of Fig. 13-9 indicates that the steel now contains 50 percent austenite and 50 percent pearlite. When the time-line reaches the right C curve, the transformation "party" is all over.

FOUR REGIONS OF TRANSFORMATION

The four regions of transformation, shown in Fig. 13-10, determine which structure the steel changes to. The regions are CP, FP, B and M. The region marked CP stands for COARSE PEARLITE or large grain pearlite. Coarse pearlite here means coarse pearlite, coarse ferrite, coarse cementite, or a combination of them. "Coarse pearlite" is an abbreviation for any of those terms. Using the abbreviation saves time in saying "combination of coarse ferrite and coarse pearlite" or "combination of coarse pearlite and coarse cementite," etc.

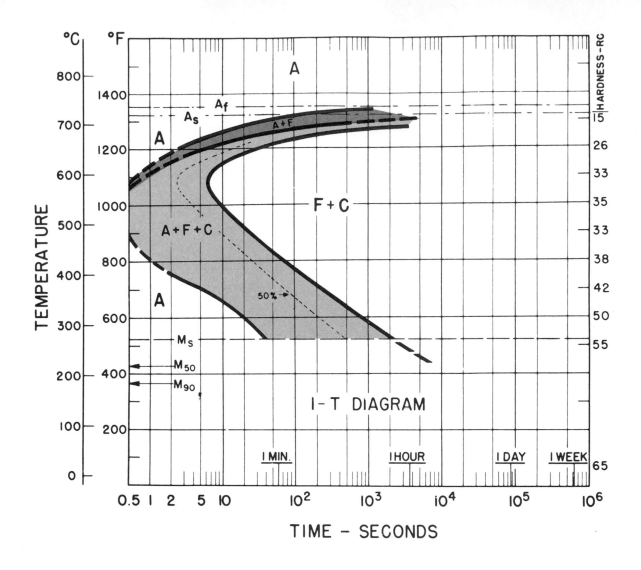

Fig. 13-6. This industrial I-T diagram is for 1060 steel.

Coarse pearlite is obtained when a steel is cooled extremely slowly. Whenever a time-line crosses this region of transformation, the steel structure will change to coarse pearlite.

FP stands for fine pearlite, fine ferrite, fine cementite, or a combination of them. FP, or FINE PEARLITE, is an abbreviation for any combination of fine ferrite, fine pearlite and fine cementite. Any time a steel time-line crosses the FP region, the steel becomes fine pearlite.

The "M" region is the MARTENSITIC REGION. If the time-line of a steel crosses the region of transformation action in the "M" region, martensite will be formed. This will occur if the steel is quenched very rapidly.

"B" stands for BAINITE. This steel structure was named after Dr. E.C. Bain, who did research on this type of steel structure.

Bainite is superior to martensite in ductility and toughness but not as good as martensite from a hardness and strength standpoint. Bainite has less ductility than fine pearlite but it is stronger and harder. Thus, bainite is a happy medium between martensite and fine pearlite.

Bainite typically reaches hardnesses of 50-55

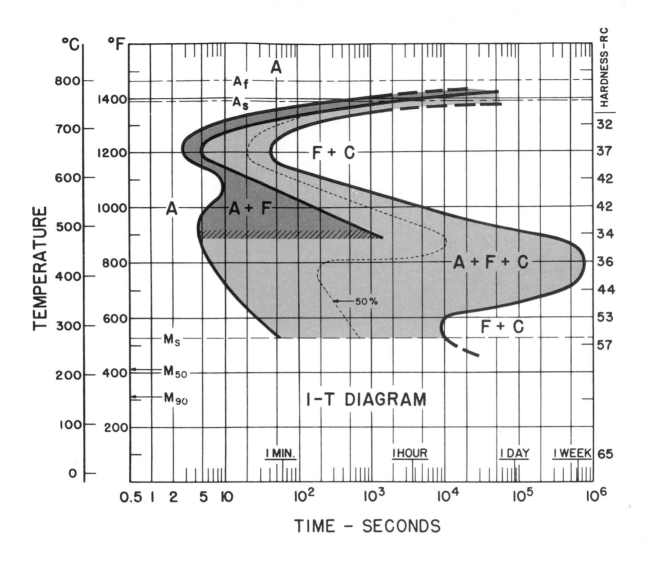

Fig. 13-7. This industrial I-T diagram is for a silicon 9261 steel.

R_C. The structural appearance of bainite when viewed under a microscope is shown in Fig. 13-11.

In summary, bainite has the advantage of fairly high strength with fairly good ductility.

EXAMPLES OF ISOTHERMAL TRANSFORMATIONS

The following examples will illustrate the use of the isothermal transformation diagram.

1. In Fig. 13-12, time-line A shows a steel that is quenched rapidly. In fact, it is quenched so rapidly that the time-line does not touch the nose of the left C curve. All transformation takes place in the "M" region. Therefore, the steel structure that is obtained is martensite.

2. Time-line B is a steel that is cooled very slowly. The time-line crosses the region of transformation action between B_1 and B_2. Therefore, this material transforms from austenite to 100 percent coarse pearlite. (Remember that coarse pearlite means some combination of coarse pearlite, coarse ferrite, and coarse cementite. It does not mean literally 100 percent pearlite.)

3. Time-line C crosses the region of transformation in the "FP" region. Therefore, steel sample C will become fine pearlite.

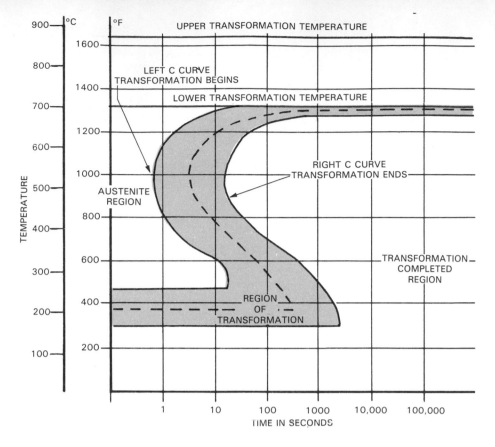

Fig. 13-8. Simplified isothermal transformation diagram shows basic graph of temperature VS time for process of cooling metal.

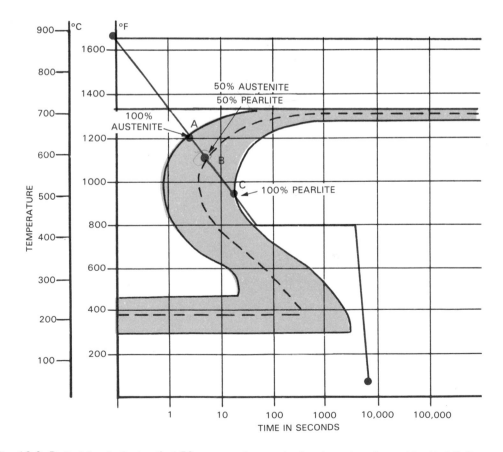

Fig. 13-9. Dotted line indicates that 50 percent of austenite has been transformed in this I-T diagram.

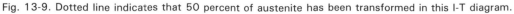

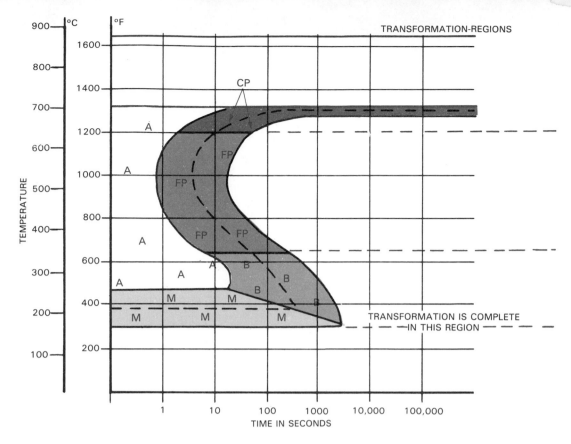

Fig. 13-10. There are four regions of transformation on an I-T diagram: CP—Coarse pearlite. FP—Fine pearlite. B—Bainite. M—Martensite.

Fig. 13-11. In this microphotograph of 1045 steel, bainite-martensite structure, bainite areas appear darker than martensitic regions.

4. Steel D transforms to bainite because the time-line crosses the region of transformation action in the "B" region.

5. In Fig. 13-13, time-line E crosses the region of transformation action in both the bainite and fine pearlite region. At point E_2, 50 percent of the austenite has transformed to fine pearlite. Then, the time-line essentially backs up and again gets closer to the left C curve. It does not move forward again until E_3. The remaining 50 percent austenite transforms to bainite between E_4 and E_5. Thus, steel E will become 50 percent fine pearlite and 50 percent bainite.

6. Example F is a more complex one. See Fig. 13-14. At F_1, steel F is still 100 percent austenite. When the time-line reaches point F_2, 25 percent of the material has transformed to fine pearlite. The time-line then moves toward the left C curve, so no further transformation takes place between F_2 and F_3. When the time-line reaches point F_4 at the

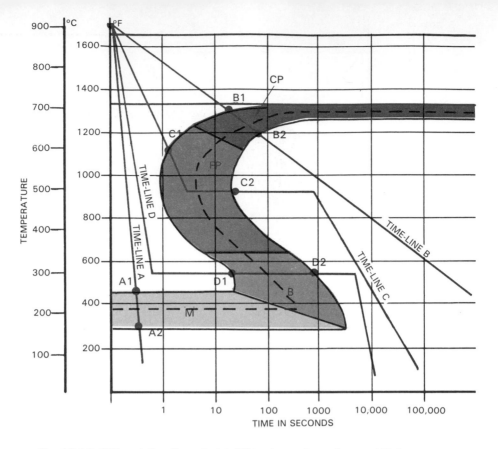

Fig. 13-12. Different time-lines depict different transformations on I-T diagrams.

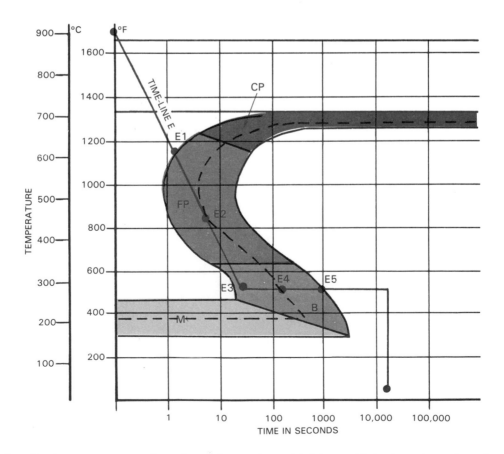

Fig. 13-13. Time-line E will cause a transformation to fine pearlite and bainite on this isothermal transformation diagram.

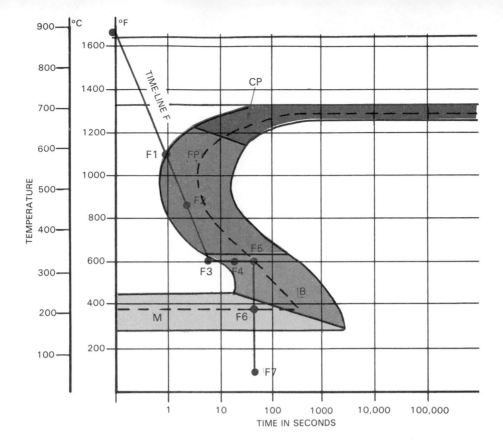

Fig. 13-14. Time-line F will cause a transformation to fine pearlite, bainite, and martensite on this isothermal transformation diagram.

25 percent mark again, transformation again takes place. Now, 25 percent more of the austenite is transformed to bainite between F_4 and F_5. At this point, the steel is cooled rapidly. No further transformation takes place until the curve again reaches the 50 percent transformation line at F_6, at which time the last 50 percent of the material becomes martensite.

Thus, steel F transforms to 25 percent fine pearlite, 25 percent bainite and 50 percent martensite.

USING INDUSTRIAL I-T DIAGRAMS

Compare the industrial I-T diagram of Fig. 13-4 to the simplified version in Fig. 13-12. These two diagrams give much of the same basic information. (Note that they are for two different types of steel.) The simplified version may be easier to read. The industrial version contains more detailed information.

In industry, I-T diagrams similar to Fig. 13-4 will be available. Hardness values for the transformed steel are shown at the right and will be discussed later in this chapter. An additional line in some industrial I-T diagrams allows you to distinguish between coarse pearlite and coarse ferrite. Figs. 13-5, 13-6, and 13-7 all have this line.

The industrial I-T diagram becomes very easy to use if you imagine a few heavy lines added, as shown in Fig. 13-15.

COMPARISON OF I-T DIAGRAMS OF DIFFERENT STEELS

Compare the I-T diagrams for three different steels in Figs. 13-16, 13-17, and 13-18. Note that the nose of the left "C" curve comes much closer to the left border in 1095 steel in Fig. 13-16 than it does in 4140 steel in Fig. 13-17. This means that 1095 steel must be quenched much faster in order to obtain martensite. The 4140 steel, which has

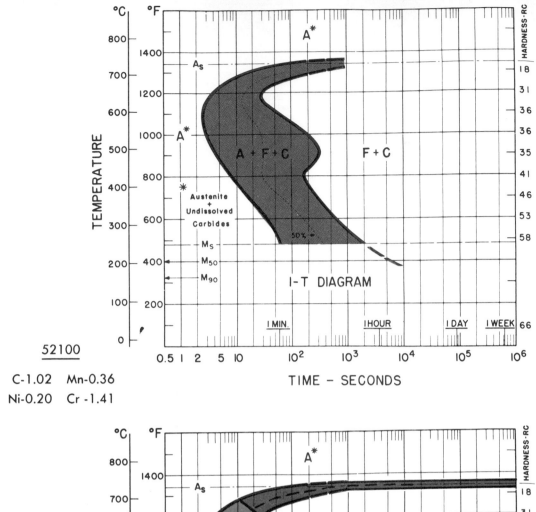

°C | °F | | | | | | | HARDNESS-RC
800 | 1400 | A*
700 | | A_S | | | | | | 18
600 | 1200 | | | | | | | 31
500 | 1000 | A* | | A + F + C | | F + C | | 36
400 | 800 | | | | | | | 36
300 | 600 | * Austenite + Undissolved Carbides | | | 50% → | | | 35
200 | 400 | M_S M_50 M_90 | | | | I-T DIAGRAM | | 41 46 53 58
100 | 200 | | | | I MIN. | I HOUR | I DAY | I WEEK | 66
0 | | 0.5 1 2 5 10 | 10^2 | 10^3 | 10^4 | 10^5 | 10^6

52100

TIME – SECONDS

C-1.02 Mn-0.36
Ni-0.20 Cr -1.41

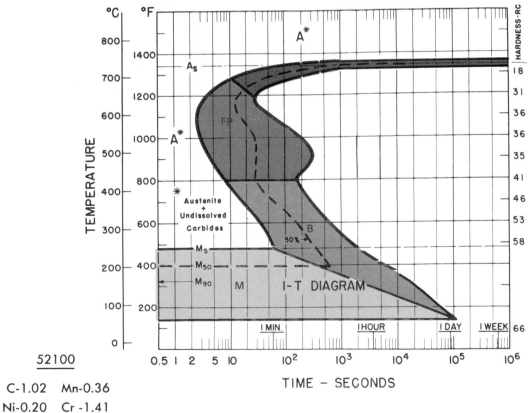

52100

TIME – SECONDS

C-1.02 Mn-0.36
Ni-0.20 Cr -1.41

Fig. 13-15. An industrial I-T diagram becomes easier to use if a few heavy lines are added.

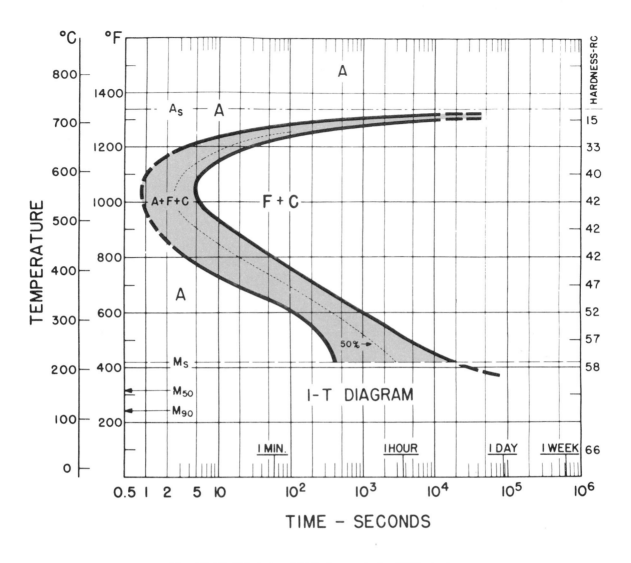

Fig. 13-16. This industrial I-T diagram is for 1095 steel.

more alloys, does not have to be quenched as rapidly to miss the nose and become martensite. Note that the 1060 steel shown in Fig. 13-18 has no space between the nose and the left border. Thus, no matter how fast the quenching speed is, 100 percent martensite cannot be obtained with 1060 steel.

Compare the shape of the C curve for 9261 steel in Fig. 13-19 with 1095 steel in Fig. 13-16, and with 4140 steel in Fig. 13-17, and with 1060 steel in Fig. 13-18. "C" curves for different steels vary considerably. Almost every alloy that is added to the steel has an effect.

PRACTICAL OBJECTIVES IN SELECTION OF STEEL

If it is important to obtain martensite, you would prefer to use a steel with a left nose that is far from the left border of the I-T diagram. For example, the 4140 steel in Fig. 13-17 is much easier to harden than the 1095 steel of Fig. 13-16.

The trick, then, in hardening is either to quench the steel very rapidly and miss the nose of the curve, or else use a more expensive, higher alloy steel whose nose has plenty of "breathing room" between it and the left border.

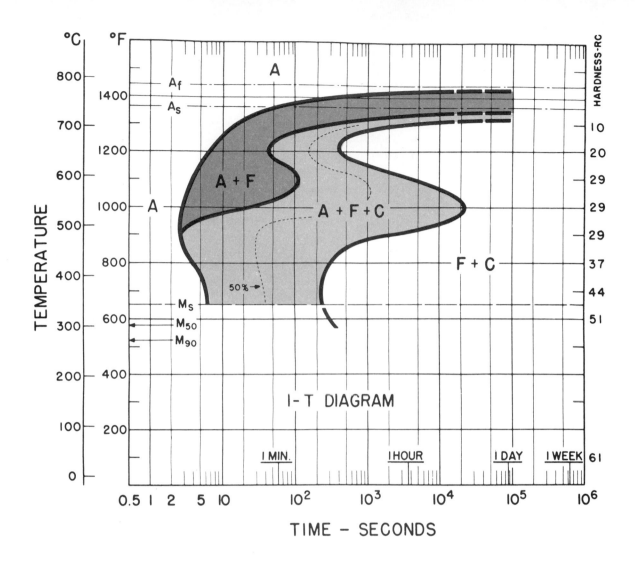

Fig. 13-17. This industrial I-T diagram is for 4140 steel.

HARDNESS

You will note that on the industrial I-T diagram, HARDNESS VALUES are given at the far right of the diagram. See Fig. 13-20. From these values and the crossing points of the time-line, you are able to determine how hard the new structure is.

An exact value of hardness generally cannot be determined, but a HARDNESS RANGE can be established. A minimum and maximum hardness value can be "picked off." The minimum possible hardness of a steel structure corresponds to the hardness at the point where the time-line

first crosses the left "C" curve. The maximum possible hardness corresponds to the value of hardness where the time-line crosses the right "C" curve.

For example, in Fig. 13-21, the hardness of sample "A" is between 18 R_C and 28 R_C. The hardness of sample "B" is between 27 R_C and 53 R_C. Sample "C" has a hardness value between 32 R_C and 66 R_C. The hardness of sample "D" is between 58 R_C and 66 R_C.

MAKING I-T DIAGRAMS

The metallurgical and steel corporations that

produce these I-T diagrams obtain their data from a series of tests: Many small, identical samples are prepared. Typical samples will be about one inch in diameter and 1/16 in. thick. Each sample is heated above the upper transformation temperature and quenched in some manner. Each sample is quenched or cooled slightly differently than the others. In order to obtain isothermal conditions, samples may be held at a given temperature by plunging the sample into a pot of hot molten lead that is maintained at 500°F (260°C), or 800°F (427°C), or 1000°F (538°C), etc.

In order to plot an I-T diagram, many samples will be used, perhaps 100 or more. One by one they are removed from the furnace, quenched, and tested for hardness.

The data that these samples generate determines the actual shape of both C curves. This data is also used to determine the hardness values at the right of the industrial diagrams.

EXAMPLES OF ISOTHERMAL TRANSFORMATIONS

The following examples illustrate several different transformations that can take place when time, temperature, and alloy content are varied.

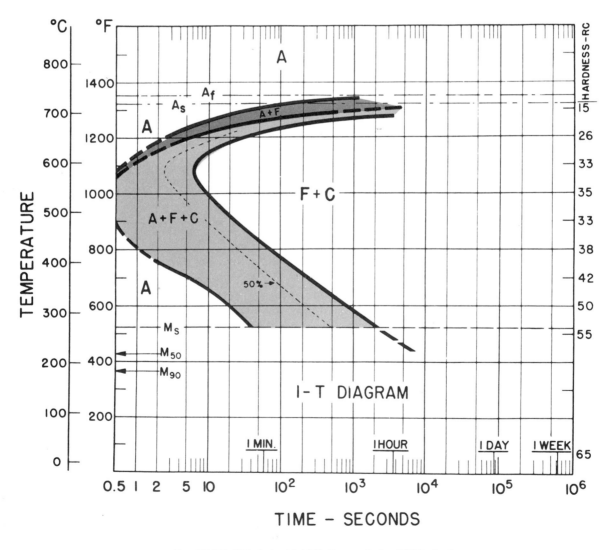

Fig. 13-18. This industrial I-T diagram is for 1060 steel.

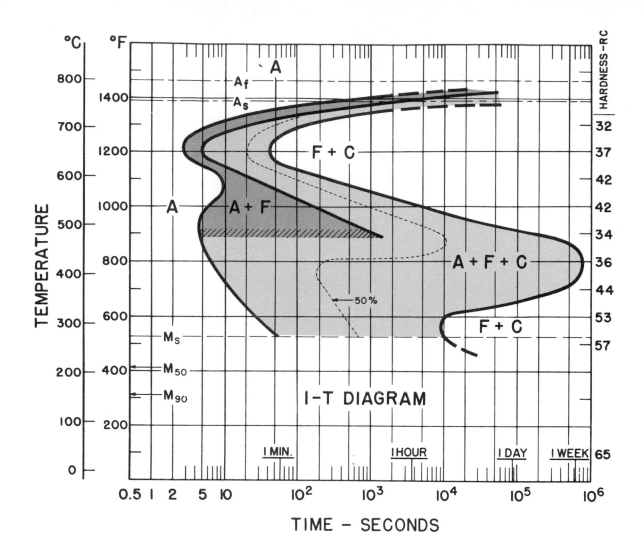

Fig. 13-19. This industrial I-T diagram is for 9261 steel.

Example A

A sample of 1340 steel is heated to 1600°F (870°C). It is then gradually cooled to 1200°F (650°C) over a 10 min. time period. Finally, it is quenched to room temperature in five sec.

The time-line for steel A is plotted in Fig. 13-22. All transformation takes place in the CP region. Therefore, transformation is to 100 percent coarse pearlite. The hardness value is below 15 R_c.

Note that quenching occurs immediately after this time-line for steel A crosses the region of transformation.

Example B

A sample of 1340 steel is heated to 1600°F and cooled to 1000°F in four hr. It is then quenched rapidly to room temperature in five sec.

The time-line for steel B is plotted in Fig. 13-23. Since all transformation takes place in the CP region, this structure becomes 100 percent coarse pearlite. The hardness value is below 15 R_c.

The same type of steel is used in Examples A and B. The difference is in the last part of the cooling cycle.

Note that this time-line crosses the region of transformation action in nearly the same place

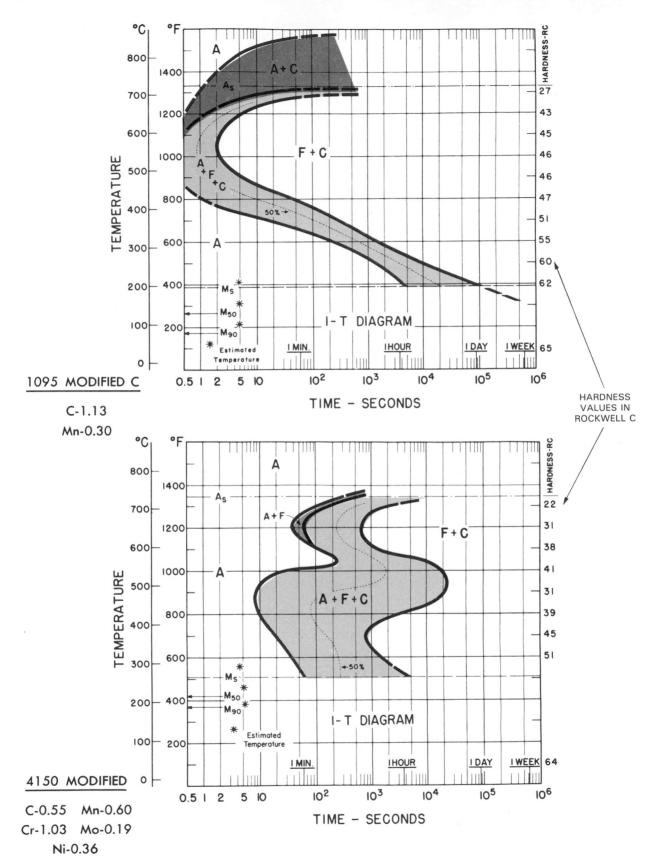

°C °F

1095 MODIFIED C

C-1.13
Mn-0.30

HARDNESS
VALUES IN
ROCKWELL C

4150 MODIFIED

C-0.55   Mn-0.60
Cr-1.03   Mo-0.19
Ni-0.36

Fig. 13-20. Hardness values on isothermal transformation diagrams are given at far right of diagram.

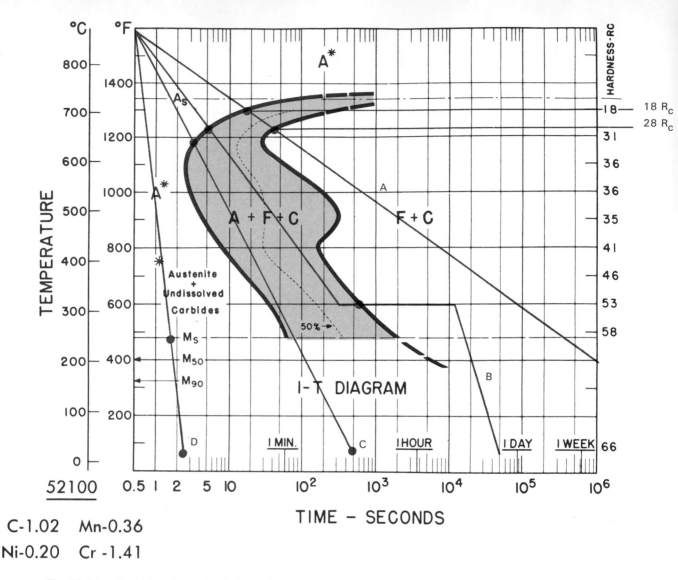

Fig. 13-21. Industrial isothermal transformation diagram for 52100 steel, with four different time-lines added.

that sample "A" does. Once the time-line has reached point B_2, theoretically, all transformation is over. Therefore, the remaining portion of the curve between B_2 and B_3 has nothing to do with determining what the final transformation will be. Sample "A" and sample "B" attain the same structure, even though sample "A" receives a quenching action immediately after transformation is over.

Example C

A sample of a silicon steel is heated to 1600°F and then quenched rapidly in water in one sec. The time-line is shown in Fig. 13-24. All transfor-

mation takes place in the "M" region. Therefore, the new structure will be 100 percent martensite. The hardness range is 57 R_c to 65 R_c.

Example D

A sample of the same silicon steel in Example C is heated to 1600°F and cooled to 600°F (316°C) in one sec. Then, it is held at 600°F for 20 sec. in a molten lead bath. Finally, it is quenched rapidly in water from 600°F down to room temperature in one sec.

The time-line for steel "D" is plotted in Fig. 13-25. All transformation takes place in the "M" region. Therefore, sample "D" will become 100 percent martensite. The hardness will be between 57 R_c and 65 R_c.

Note that sample "C" and sample "D" both end up at 100 percent martensite and both end up with the same hardness values even though sample "D" was permitted more time to cool. There is an advantage to cooling in this manner. This will be discussed in Chapter 14. This technique is known as "martempering."

Example E

A sample of 1095 steel is heated to 1600°F. It is then cooled to 600°F in two sec. Then, it is held at 600°F for one hr. Finally, it is quenched rapidly in water in one sec. until it reaches room temperature.

The time-line for steel E is plotted in Fig. 13-26. Note that the time-line slightly crosses the nose of the left C curve. Because of this, approximately 10 percent fine pearlite is formed. Between points E_1 and E_2, the time-line leaves the region of transformation action and reenters the austenitic region. Even though the time-line is in the austenitic region, the 10 percent fine pearlite will not transform back to austenite. Further transformation waits until the time-line reaches point E_3 (10 percent point). Then, the remainder transforms to bainite. Thus, sample E becomes 10 percent fine pearlite and 90 percent bainite. The hardness range is between 40 R_c and 52 R_c.

Example F

A sample of 1095 steel is heated to 1600°F and quenched rapidly to 1000°F in one sec. It is held at 1000°F for two more sec. and then quenched

to room temperature in one additional sec.

The time-line for steel F is shown in Fig. 13-27. Transformation takes place in both the "FP" region and the "M" region. In the "FP" region, 50 percent of the austenite is transformed to fine pearlite. The time-line at point F_2 extends 50 percent of the way across the region of transformation action. As the time-line moves from point F_2 to point F_3, no additional transformation takes place because the time-line is essentially moving backwards toward the left C curve and the austenite region. Further transformation takes place at point F_3 when the time-line again reaches the 50 percent line. Thus, the final structure will be 50 percent fine pearlite and 50 percent martensite. The hardness will be between 41 R_c and 66 R_c.

Example G

A sample of 1566 steel is heated to 1600°F and cooled to 950°F (510°C) in eight sec. It is then cooled to 600°F in 12 more sec. (20 sec. total). It is held at 600°F for seven more min. and then quenched rapidly to room temperature in water.

The time-line for sample G is shown in Fig. 13-28. Transformation will take place between points G_1 and G_2 in the "FP" region, between points G_3 and G_4 in the bainite region, and finally between points G_5 and G_6 in the martensitic region. The final result will be 50 percent fine pearlite, about 20 percent bainite and 30 percent martensite. The 50 percent fine pearlite is not lost as the time-line backs up toward the left C curve and the austenitic region. Note that the martensitic transformation does not begin until it reaches point G_5 since 70 percent of the material has already been transformed.

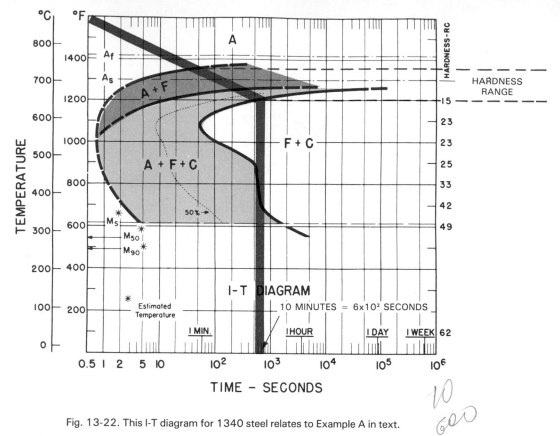

Fig. 13-22. This I-T diagram for 1340 steel relates to Example A in text.

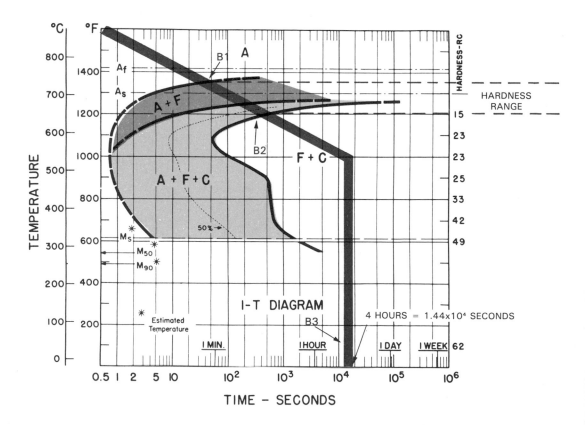

Fig. 13-23. This I-T diagram for 1340 steel relates to Example B in text.

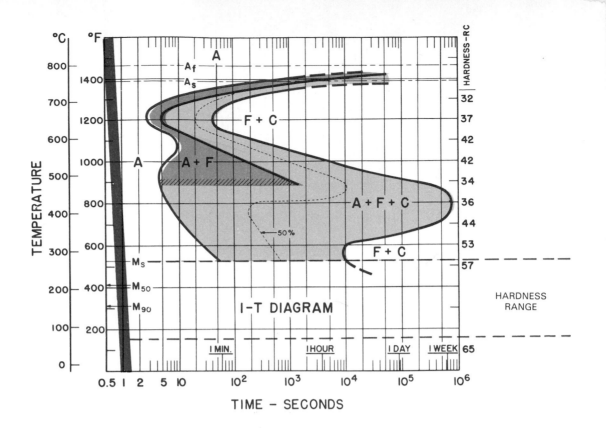

Fig. 13-24. This I-T diagram for a silicon steel relates to Example C in text.

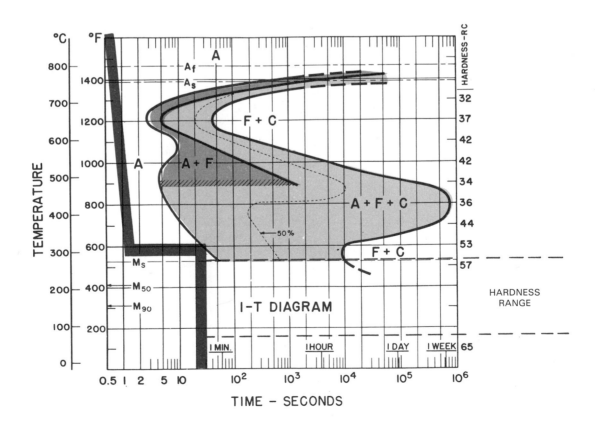

Fig. 13-25. This I-T diagram for a silicon steel relates to Example D in text.

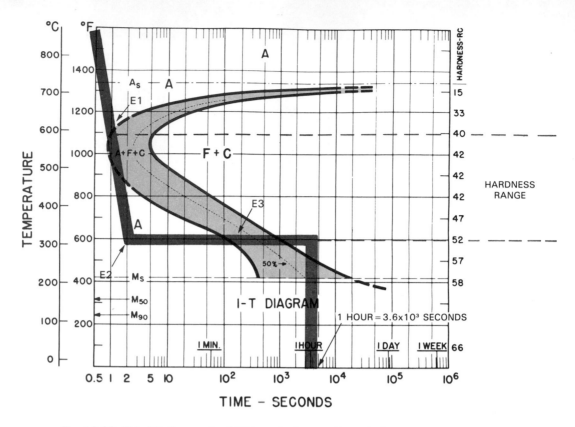

Fig. 13-26. This I-T diagram for 1095 steel relates to Example E in text.

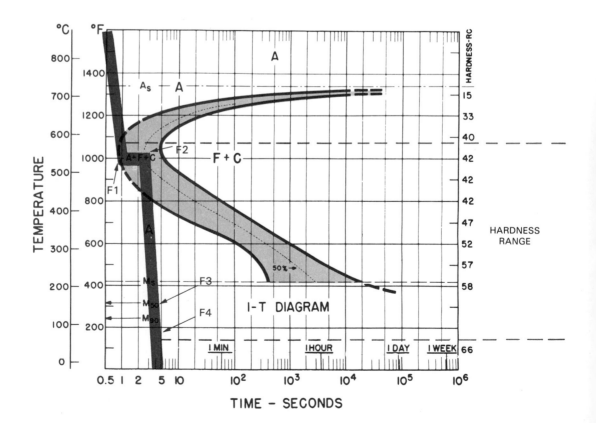

Fig. 13-27. This I-T diagram for 1095 steel relates to Example F in text.

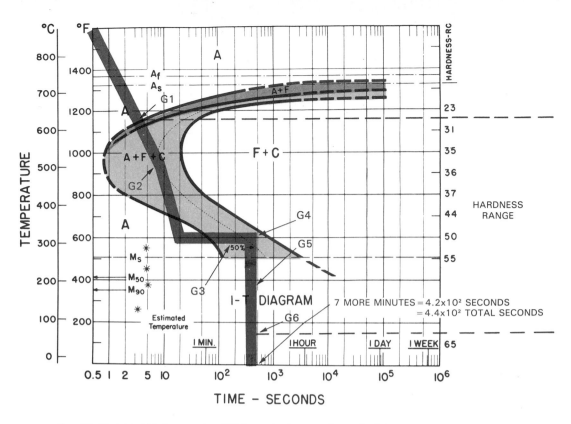

°C °F

HARDNESS-RC

800 ── 1400   A   Af As   G1
700 ──   A+F
          1200   A
600 ──
          1000   A+F+C   F+C   23
500 ──   G2   31
400 ──   800   G4   35
          A   36
300 ──   600   G5   37
              Ms   50%   44
200 ──   400   M50   50
              M90   G3   55
100 ──   200   Estimated   I-T DIAGRAM   7 MORE MINUTES = 4.2x10² SECONDS
              Temperature   G6   = 4.4x10² TOTAL SECONDS
  0 ──   1 MIN.   1 HOUR   1 DAY   1 WEEK 65

HARDNESS RANGE

TEMPERATURE

0.5 1 2 5 10   10²   10³   10⁴   10⁵   10⁶

TIME – SECONDS

Fig. 13-28. This I-T diagram for 1566 steel relates to Example G in text.

TEST YOUR KNOWLEDGE

Write your answers on a separate sheet of paper. Do not write in this book.

1. What is another name for an I-T diagram?
2. What valuable piece of information does the iron-carbon phase diagram lack?
3. On any I-T diagram, what steel structure corresponds to a time-line that has not yet crossed either C curve?
   a. Fine pearlite.   c. Austenite.
   b. Coarse pearlite.   d. Martensite.
4. Name the two transformation regions that correspond to the letters M and FP.
5. CP is an abbreviation that stands for not only coarse pearlite but also for _____ and _____ .
6. Which has better ductility, bainite or martensite?
7. Which has higher hardness, bainite or martensite?

For questions 8 to 14, use the appropriate industrial isothermal transformation diagram. Determine the structure that would result from the following quenching cycles:

8. Type of steel is 4140 (Fig. 13-5).
   First, heat to 1600°F.
   Then, cool to 1000°F in two sec.
   Then, hold at 1000°F for 20 min.
   Then, cool to room temperature in two sec.
   Which structure would this produce?
   a. 100 percent martensite.
   b. 100 percent bainite.
   c. 100 percent fine pearlite.
9. Type of steel is 1095 (Fig. 13-4).
   First, heat to 1500°F.
   Then, cool to 500°F in one sec.
   Then, hold at 500°F for 10 hr.
   Then, cool to room temperature in one hr.
   Which structure would this produce?
   a. 100 percent martensite.
   b. 100 percent bainite.
   c. 100 percent fine pearlite.

10. Type of steel is 52100 (Fig. 13-15).
    First, heat to 1500°F.
    Then, cool to 500°F in one sec.
    Then, hold at 500°F for 20 sec.
    Then, cool to room temperature in two sec.
    Which structure would this produce?
    a. 100 percent martensite.
    b. 100 percent bainite.
    c. 100 percent coarse pearlite.

11. Type of steel is 52100 (Fig. 13-15).
    First, heat to 1500°F.
    Then, cool to 1300°F in two sec.
    Then, hold at 1300°F for 60 sec.
    Then, cool to room temperature in two sec.
    Which structure would this produce?
    a. 100 percent bainite.
    b. 100 percent fine pearlite.
    c. 100 percent coarse pearlite.
    d. 100 percent martensite.

12. Type of steel is 9261 (Fig. 13-7).
    First, heat to 1500°F.
    Then, cool to 700°F in one sec.
    Then, hold at 700°F for 30 sec.
    Then, cool to room temperature in two sec.
    Which structure would this produce?
    a. 50 percent each of fine and coarse pearlite.
    b. 50 percent each of fine pearlite and bainite.
    c. 50 percent each of bainite and martensite.

13. Type of steel is 1095 (Fig. 13-4).
    First, heat to 1500°F.
    Then, cool to 1000°F in two sec.
    Then, hold at 1000°F for one sec.

Then, cool to 200°F in seven sec.
Which structure would this produce?
a. 50 percent each of coarse and fine pearlite.
b. 50 percent each of fine pearlite and martensite.
c. 50 percent each of fine pearlite and bainite.

14. Type of steel is 1060 (Fig. 13-6).
    First, heat to 1450°F.
    Then, cool to 450°F in five sec.
    Then, hold at 450°F for 10 sec.
    Then, cool to room temperature in three hr.
    Which structure would this produce?
    a. 50 percent fine pearlite and 50 percent martensite.
    b. 30 percent fine pearlite and 70 percent martensite.
    c. 100 percent fine pearlite.
    d. 100 percent martensite.

15. Determine the value of hardness range that the 4140 steel in question 8 will reach.

16. Determine the value of hardness range that the 1095 steel in question 9 will reach.

17. Determine the value of hardness range that the 52100 steel in question 10 will reach.

18. Determine the value of hardness range that the 52100 steel in question 11 will reach.

19. Determine the value of hardness range that the 9261 steel in question 12 will reach.

20. Determine the value of hardness range that the 1095 steel in question 13 will reach.

21. Determine the value of hardness range that the 1060 steel in question 14 will reach.

14   TEMPERING

After studying this chapter, you will be able to:

☐ Explain what tempering is.
☐ State the purpose of tempering and when it should be done.
☐ Point out how tempering affects distortion and hardness.
☐ Discuss some of the practical aspects of tempering.
☐ Describe the three special types of tempering.

WHAT IS TEMPERING?

TEMPERING is defined as "the process of reheating steel after hardening to a temperature which is below the lower transformation temperature, followed by any rate of cooling, for the purpose of increasing the ductility and toughness of the steel."

We need to expand on this definition and put it into layman's terms. After steel has been quenched, it may be hard and brittle. There are internal stresses in the steel. There is a good chance of distortion or cracking taking place. If the steel is immediately reheated after it is quenched, the internal stresses will relax. This reheating temperature does not have to be extremely high. The metal is normally given a quick "stress relieving," at a medium temperature, perhaps about 800°F (427°C).

Tempering takes place only after quenching and hardening have previously been completed. Tempering consists of heating the steel to a temperature that is not extremely high, but just high enough to remove some of the internal stresses.

Tempering is sometimes called DRAWING. The term "drawing" apparently comes from the conception that, in tempering, the hardness is "drawn back."

PURPOSE OF TEMPERING

There is a popular misconception of tempering. Many people think that tempering hardens steel. This is false. Tempering actually softens steel.

The purpose of tempering is to remove brittleness and to reduce internal stresses. Tempering reduces the chance of distortion. Tempering reduces the likelihood of cracks. Unfortunately, tempering also reduces the hardness and the strength. However, the amount of strength and hardness that is lost is not excessive, and the benefits of eliminating the brittleness and internal stresses generally offset the problems caused by the slight loss of hardness and strength. By removing these internal stresses: the material is more ductile; has better toughness; has better impact resistance; and is more easily machined and cold worked.

EFFECTS OF TEMPERING

To further study the process of tempering, compare tempering to normalizing and annealing. With tempering, the metal property changes are less severe. There is less softening and less strength loss than in normalizing or annealing.

Tempering is somewhat similar to process annealing. But tempering is faster. Tempering causes less change in the metal. Tempering takes less time.

In summary, tempering is a quick heat-treating process that does not remove all of the internal stresses and brittleness characteristics, but removes enough of them that it is worth doing in industry.

Fig. 14-1 summarizes the effects of tempering on hardness, strength, toughness, brittleness, ductility, internal stresses, distortion, cracking, machinability, and formability.

EFFECTS OF TEMPERING	
HARDNESS	Decreased
STRENGTH	Decreased
TOUGHNESS	Increased
BRITTLENESS	Decreased
DUCTILITY	Increased
INTERNAL STRESS	Decreased
DISTORTION	Reduced
CRACKING	Reduced
MACHINABILITY	Improved
FORMABILITY	Improved

Fig. 14-1. Chart gives effects of tempering on material properties.

PRACTICAL ASPECTS OF TEMPERING

At what temperature should tempering be done? Tempering is generally done between 300° − 1200°F (149° − 649°C), Fig. 14-2. If the metal is heated less than 300°F, essentially no change takes place and tempering becomes a waste of time. Over 1200°F, structural changes are ready to start taking place. This is not the purpose of tempering. Also, to heat it over 1200°F, involves more time than is necessary.

Materials that are tempered near 1200°F have more property changes than those that are heated in the 300° − 500°F (149° − 260°C) range. Fig. 14-3 shows the effects of different tempering temperatures. Note that the hardness change is

considerably more when the metal is tempered at 1000°F (538°C) than when it is tempered at 400°F (204°C).

How soon should tempering be started after quenching is completed? Tempering should be done immediately after quenching. There is no benefit in waiting. If too much time elapses between quenching and tempering, the internal stresses have time to work on the metal. Then, cracking or distortion can occur before tempering is started.

How long should the metal be left at the elevated temperature during tempering? In Fig. 14-4, the effects of soaking the metal at the tempering temperature for a long period of time are apparent. Note that when metal is heated to 1000°F, the effect of the first 30 min. of soaking time is very evident. After the first 30 min., the softening effects are less dramatic unless the soaking time involves "days" instead of

Fig. 14-2. This tempering furnace employs molten salt and can be externally heated by gas or oil-fired burners. It uses a steel pot which is well insulated with brick. (Ajax Electric Company)

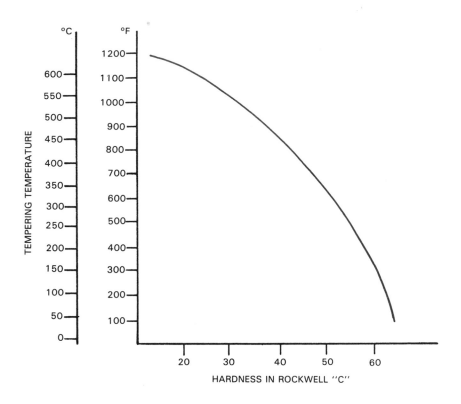

Fig. 14-3. Graph shows effects of different tempering temperatures on hardness of a metal.

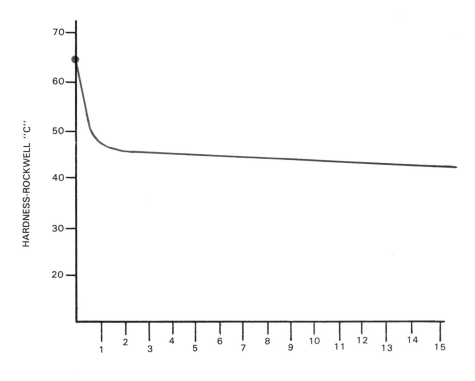

SOAKING TIME IN "HOURS"

Fig. 14-4. Graph reveals effects of soaking time at tempering temperature on hardness of a metal.

"minutes." Steel that is left to soak in the oven for a day, instead of an hour, softens about five units Rockwell "C."

Is tempering always desirable? No, tempering is not always desirable. When hardness is important in a steel, tempering may become a problem. Tool bits can get hot during cutting. The metal may reach a tempering temperature. This tempering action can cause a slight softening in the tool bit.

Undesired tempering can also occur in welding and can cause softening of the metal. In mechanisms of machines, tempering can result from two surfaces rubbing together and building up frictional heat. In these cases, tempering is done unintentionally and may cause metal to lose hardness and strength.

SPECIAL TYPES OF TEMPERING

There are three special tempering techniques which are often used in metallurgy to obtain special structures. They reduce the tendency to distort or crack even more than regular tempering. These special techniques are known as:
1. Martempering.
2. Austempering.
3. Isothermal quenching and tempering.

These three special techniques are compared to regular quenching and tempering in Figs. 14-5, 14-6, 14-7, and 14-8.

GENERAL QUENCHING AND TEMPERING

Fig. 14-5 shows the general method of quenching and tempering. Note that the time-line

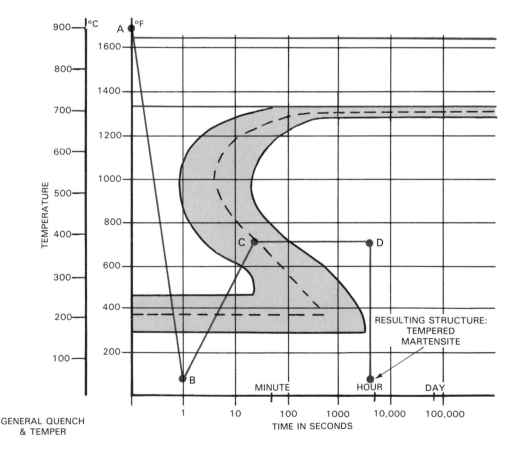

Fig. 14-5. Diagram has time-line superimposed for general quenching and tempering.

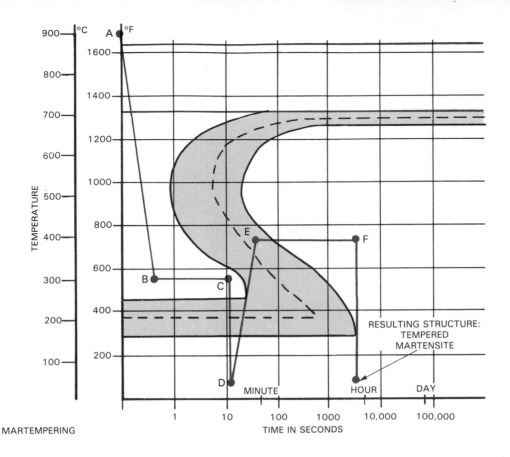

Fig. 14-6. Time-line is given for martempering.

curve misses the nose of the C curve on the isothermal transformation diagram. The time-line then crosses the martensite transformation region and the steel becomes 100 percent martensite. At this point, quenching is completed and tempering is started. As quickly as possible, the material is reheated (line B-C in Fig. 14-5) to a tempering temperature in the range of 500°–1000°F (260°–538°C). The steel is soaked at that temperature for several hours and then quenched back to room temperature in a short period of time.

"Tempered martensite" is the name given to the resulting structure. It is very similar to martensite, but its properties are drawn back slightly. Thus, compared to martensite, this structure is slightly less hard, less strong, and less brittle; but slightly more ductile, tougher, more stress-free, more distortion-free, and more crack-free.

Note that the isothermal C curves in Figs. 14-5, 14-6, 14-7, and 14-8 pertain only to the quenching action. Once the material is quenched, the tempering time-line has nothing to do with the "C" curves that it is superimposed on in these diagrams.

MARTEMPERING

MARTEMPERING is very similar to general tempering. The quenching action is a little more gentle. In Fig. 14-6, note that after missing the nose of the C curve, the material is held at about 500°–600°F (260°–316°C) for a short period of time. Because of this, the quench is not quite as drastic when it finally crosses the martensitic transformation region. After quenching, the material is 100 percent martensite, but is slightly more stress-free than the 100 percent martensite that was attained from general quench and temper. After quenching to 100 percent martensite, the steel is immediately tempered. It is reheated to an intermediate temperature, held there for a period of time, and requenched back to room temperature.

The structure that results from martempering also is tempered martensite. The advantage of martempering over general tempering is the reduced tendency toward cracking and distortion. The disadvantage of martempering versus general tempering is that leveling out at 500°F is somewhat time consuming and inconvenient. Therefore, it is done only when the tendency to distort or crack is very critical.

AUSTEMPERING

AUSTEMPERING, Fig. 14-7, starts out like martempering and general tempering. It misses the nose of the "C" curve. However, when the material that is being austempered reaches the leveling-out temperature of 500−600°F, it is soaked there for a long period of time. The time period is sufficiently long for structural change to take place in the bainite region. Thus, marten-

site is never formed. After the bainite structure evolves, the steel is held at that temperature for a period of time and then quenched to room temperature.

Austempering is a much more gentle heat-treating technique than either martempering or general tempering. The structure that results will have less internal stress, less distortion tendencies, and less cracking tendencies. Also, it will *not* be as hard or strong as martempering or general tempering.

Austempering is generally limited to thin parts wherein distortion can be a serious problem. Springs, lockwashers, needles, and other fine parts are often austempered.

Austempered metals are usually tougher and more ductile than martempered or general

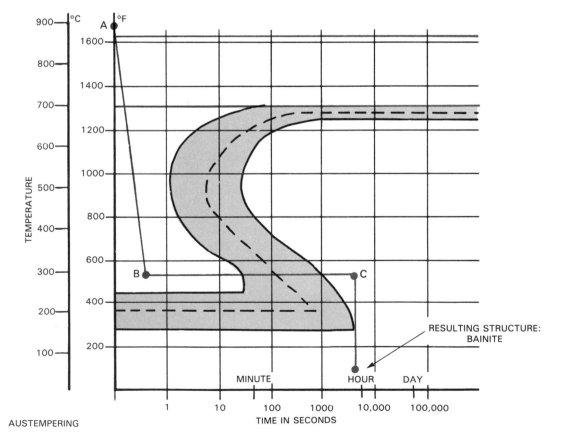

Fig. 14-7. Time-line is shown for austempering.

tempered metals. They have better impact resistance since they contain no hard-brittle martensite.

ISOTHERMAL QUENCHING AND TEMPERING

ISOTHERMAL QUENCHING AND TEMPERING is a happy medium between martempering and austempering. It produces a material that is harder and stronger than austempering. It produces a material that is more ductile and stress-free than martempering. The structure that results is a combination of bainite and tempered martensite.

The isothermal quenching and tempering process goes like this: When quenching is about 50 percent completed, the steel is held at constant temperature in a semi-transformed condition, see Fig. 14-8. This temperature is about 300° − 400°F

(149° − 204°C). Between A and B, about half of the steel transforms from austenite to martensite, and about half of the steel remains in the austenitic condition.

After holding the steel for a few seconds at this intermediate temperature, it is heated to a higher tempering temperature. The remaining austenite changes to bainite. As the steel soaks at the tempering temperature for a period of time, many internal stresses are removed. Finally, the steel is cooled to room temperature.

COMPARISON OF PROPERTIES OBTAINED IN TEMPERING

At this time, a chart would be helpful in comparing steel properties obtained in general tempering, martempering, austempering, and isothermal quenching and tempering. Fig. 14-9

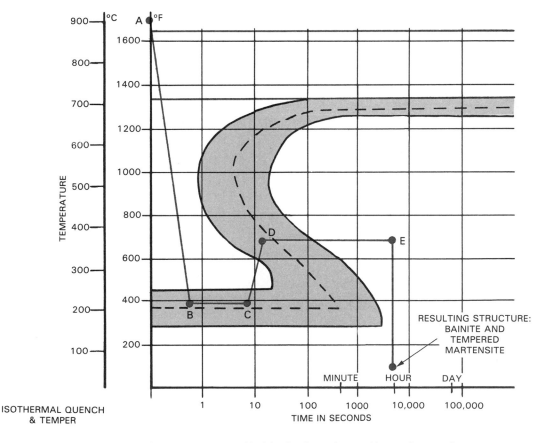

Fig. 14-8. Time-line is provided for isothermal quenching and tempering.

COMPARISON OF TEMPERING METHODS

	General Tempering	Martempering	Austempering	Isothermal Quench & Tempering
HARDNESS	Hardest	2nd	4th	3rd
STRENGTH	Strongest	2nd	4th	3rd
BRITTLENESS	Most Brittle	2nd Most	Least Brittle	2nd Least
DUCTILITY	4th	3rd	Most Ductile	2nd
TOUGHNESS	4th	3rd	Toughest	2nd
INTERNAL STRESS	Most Internal Stress	2nd Most	Least Internal Stress	2nd Least
DISTORTION	Most Distortion	2nd Most	Least Distortion	2nd Least
CRACKING	Most Cracking	2nd Most	Least Cracking	2nd Least
MACHINABILITY	4th	3rd	Easiest	2nd
FORMABILITY	4th	3rd	Easiest	2nd
CRYSTAL STRUCTURE	Tempered Martensite	Tempered Martensite	Bainite	Bainite and Tempered Martensite
COMPARATIVE SEVERITY	Most Drastic	2nd	4th	3rd

Fig. 14-9. Chart compares properties obtained with different tempering methods.

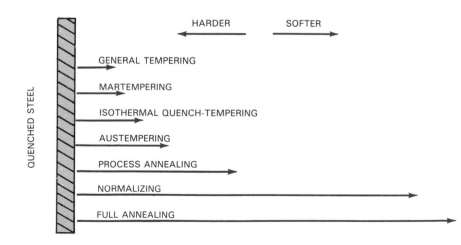

Fig. 14-10. Graph compares effects of different heat-treating techniques on hardness.

compares the effects of these four different tempering techniques regarding hardness, strength, brittleness, ductility, toughness, internal stresses, distortion, cracking tendencies, and crystal structure. Note that general tempering is the least gentle of the four.

COMPARISON OF HEAT-TREATING TECHNIQUES

Fig. 14-10 is a grand chart which compares the effects of all types of tempering and all types of annealing and normalizing.

Note that all of these secondary heat-treating operations cause the quenched material to become softer. Tempering will have the least effect because it is the least gentle of the secondary heat-treating methods. Annealing in a temperature controlled furnace is the most gentle.

TEST YOUR KNOWLEDGE

Write your answers on a separate sheet of paper. Do not write in this book.

1. Does tempering harden metal or soften it?
2. What does "drawing" mean as the term applies to heat treating?
3. What is the purpose of tempering?
4. Which of the following properties does tempering normally improve?
   a. Strength.   d. Shock resistance.
   b. Hardness.   e. Toughness.
   c. Ductility.
5. How long should one normally wait after a sample has been quenched before beginning tempering?
6. Which of the following temperatures would be a typical tempering temperature for steel?

a. 100°F.   c. 1350°F.
b. 500°F.   d. 1700°F.

7. What happens if a metal is soaked at the tempering temperature for a long period of time?
8. When is tempering not desirable?
9. What is the name given to the special tempering process wherein a metal is quenched to a temperature of about 500° or 600°F, held there for a few seconds, quenched to martensite, and then tempered?
10. What is the special name given to the special tempering process wherein a metal is quenched to a temperature of about 500° or 600°F, and held at that temperature until bainite is formed?
11. Which of the following four processes is the most drastic and would be the most likely to produce cracking or distortion?
    a. General quenching and tempering.
    b. Martempering.
    c. Austempering.
    d. Isothermal quenching and tempering.
12. Which of the following four processes is the most gentle and would be the least likely to produce cracking or distortion?
    a. General quenching and tempering.
    b. Martempering.
    c. Austempering.
    d. Isothermal quenching and tempering.
13. Which of the following heat-treating processes would cause the most softening to take place in a material?
    a. Full annealing.
    b. Normalizing.
    c. Tempering.
14. How many of the following three processes tend to soften a material and make it more ductile?
    a. Annealing.
    b. Normalizing.
    c. Tempering.

This research and development facility located in Salem, Ohio, generates many studies, developments, and processes related to the manufacture of heat-treating equipment. Studies can include: energy conservation, new approaches to solving heat-treating problems, new heat-treating methods, experimental work to determine the best atmospheres, production rates, and processes; determinations of heating and cooling times and curves for furnace design; the evaluation of the life of radiant tubes, burners, and electric heating elements used in furnace construction. (The Electric Furnace Co.)

15 SURFACE HARDENING

After studying this chapter, you will be able to:

☐ Explain what surface hardening and case hardening are.
☐ Use practical examples to show why surface hardening is a useful metallurgical process.
☐ Describe the three basic methods of surface hardening.
☐ Identify eight different industrial processes — like nitriding, pack carburizing, and flame hardening — that are used to surface-harden materials.
☐ Compare the advantages and disadvantages of the eight different processes.

SURFACE HARDENING/CASE HARDENING

Another name for "surface hardening" is "case hardening." Although there is a slight technical difference in their meanings, we will consider them to mean exactly the same thing.

SURFACE HARDENING can be defined as "a heat-treating process which creates a thin, hard, wear-resistant layer on the outer surface of a material, while permitting the inner core of the material to remain soft and ductile."

A case-hardened part is like an apple, a loaf of bread, or a tough plastic bag full of softer grains of wheat. See Fig. 15-1. In each case, a harder, stronger, outer skin or shell protects a softer interior. Of course, when we talk about metals, both the outer shell and the inside are much harder than these examples, but the situation is the same. A "hard case" covers a "softer inside."

Fig. 15-1. There are common products with same characteristics as case-hardened parts.

SURFACE HARDENING IN STEEL

In steel, the hard case usually is martensite, Fig. 15-2. The softer interior is a more ductile form of steel such as ferrite, pearlite, cementite, or a combination of these steel structures. Often, this strange mixture in a metal is very desirable. Here are some typical examples of when this is true.

CASE-HARDENED RATCHET WHEEL

The ratchet wheel shown in Fig. 15-3 rotates at a constant speed. A cam wheel rides on the teeth of the ratchet wheel. A strong spring keeps the cam wheel in contact with the ratchet wheel. After each ratchet tooth lifts the cam wheel, the cam arm actuates a switch. This indirectly measures the rotational speed of the ratchet wheel.

However, there can be a problem. After many hours of running, the cam wheel pressure will wear off the tips of the ratchet wheel teeth, Fig. 15-4, as well as some of the surface of the wheel itself. Unless the ratchet wheel has a very hard, wear-resistant surface, it will live a very short life.

It would *not* be desirable to make the entire ratchet wheel out of a hard material like martensite. Martensite would be too brittle to withstand the shock of the repeated blows as the cam wheel falls from the tip of the tooth back down to the wheel surface. The ratchet wheel would probably crack. Therefore, it *is* desirable for the ratchet wheel to have a soft interior to withstand the shock, and a hard outer case to resist wear.

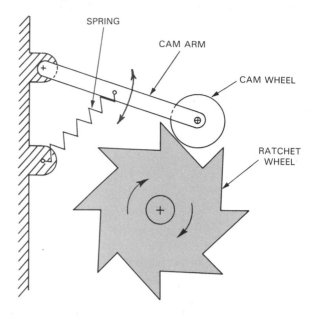

Fig. 15-3. Case hardening helps a ratchet wheel stand up to many hours of surface contact.

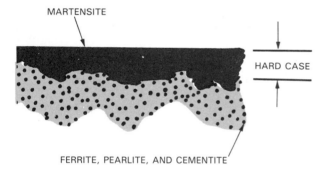
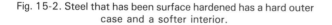

Fig. 15-2. Steel that has been surface hardened has a hard outer case and a softer interior.

OTHER EXAMPLES OF THE NEED FOR CASE HARDENING

Some cutting tools, Fig. 15-5, need a hard outside cutting surface, but will fracture easily if the entire tool is hard and brittle. Bearings, piston pins, crankshafts, and cams all must stand up to constant rubbing and metal wear, while resisting a variable amount of shock loading.

One of the first known examples of case hardening occurred in B.C., when warriors used crude methods to surface harden the tips of

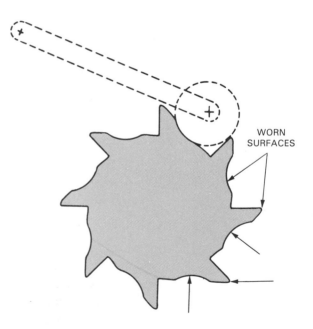

Fig. 15-4. Pressure from cam wheel can cause wear on a ratchet wheel.

their spears. They were able to pierce the enemies' armor with their case-hardened tips without cracking off the points.

In all of these examples, a hard martensitic case is desired. In all of these examples, a soft interior

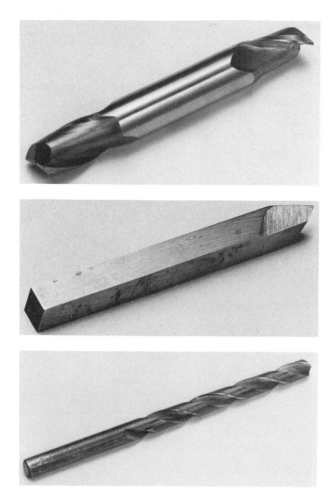

Fig. 15-5. Case hardening is used sometimes to help cutting tools last longer.

is needed. Case hardening (surface hardening) is the answer.

THICKNESS OF CASE HARDENING

The skin of case-hardened parts may be as thick as 1/8 in. or thinner than .001 in. See Fig. 15-6.

THREE BASIC METHODS OF CASE HARDENING

There are three basic methods of adding a hard case:
1. Carbon or Carburizing Method.
2. Nitrogen Method.
3. Localized Heating Method.

.001" TO 1/8"

Fig. 15-6. A typical thickness range for case-hardened parts.

Today, eight common industrial processes employ these three methods of case hardening.

THE CARBON OR CARBURIZING METHOD

In the carbon method of surface hardening, considerable amounts of carbon are crowded into the outer surface of the steel. Low carbon steel, in itself, is not capable of being appreciably hardened by quenching. The addition of carbon to it increases its ability to be hardened. After the surface of the part is impregnated with carbon and the part is heated above the transformation temperature, it is quenched. The outside, carbon-enriched surface then becomes hard.

THE NITROGEN METHOD

In the nitrogen method of surface hardening, many atoms of nitrogen are added to the outer shell of the material in much the same way that carbon is added in carburizing. The nitrogen chemically combines with its neighboring steel atoms and forms extremely hard nitride compounds.

THE LOCALIZED HEATING METHOD

In the localized heating method, no additional elements are added. The higher carbon steel used is capable of being hardened by heating and quenching. Heat is applied to only the surface of the metal. Then the entire part is quenched, but only the surface becomes hard. The core of the part remains unchanged because it was never heated over its transformation temperature. The key trick here is, *"How can you heat the outside skin of a metal to red hot, without heating the interior as well?"* Two popular surface-hardening processes — flame hardening and induction hardening — can do this.

EIGHT SURFACE-HARDENING PROCESSES

The following eight processes are among the most common surface-hardening techniques today. *Can you guess from their names which three use the Carbon Method? Which one uses the Nitrogen Method? Which two use the Localized Heating Method? Which two use a combination of the Carbon Method and the Nitrogen Method?*

1. Pack carburizing
2. Gas curburizing
3. Liquid carburizing
4. Nitriding
5. Carbo-nitriding
6. Cyaniding
7. Flame hardening
8. Induction hardening

CARBURIZING IN GENERAL

Carburizing, or the Carbon Method, is the oldest of the three methods of surface hardening. Pack carburizing is the oldest case-hardening process of the eight. Today, carburizing is still the most widely used case-hardening technique. It is also the least expensive, overall.

The technique of CARBURIZING involves bringing a carbon-rich material into contact with the steel which is to be case hardened. The carbon atoms are enticed to move from the carbon-rich material into the outer surface of the steel parts. This forms a carbon-rich layer or "case" which is capable of being heat treated to a very hard condition.

It should be emphasized that the addition of carbon to the surface does not cause hardening directly. However, after carburizing, a secondary heat-treating step is employed to cause the hardening action. The opportunity to stall the hardening mechanism until later is an advantage

of the carburizing method that will be discussed later in this chapter.

PACK CARBURIZING

In PACK CARBURIZING, the parts to be surface hardened are "packed" into a metal basket or pan along with a carbonaceous material. See Fig. 15-7. A CARBONACEOUS MATERIAL is a material that contains much carbon and is willing to give up the carbon when heated. This carbonaceous material completely buries or surrounds the parts and makes physical contact with each part.

Fig. 15-7. In pack carburizing, parts are loaded into a metal box filled with a carbonaceous material.

The carbonaceous material used for pack carburizing may be most any form of carbon. Most commonly, it is either charcoal or coke. Other materials frequently used include: bone, shells, peach pits, leather, beans, coal, nuts, or hardwood.

The entire pan of parts is heated in a furnace that is hot enough to convert some of the carbon to carbon monoxide. The carbon monoxide penetrates the surface of the parts, where it is ab-

sorbed by the austenite and thus deposits a thin shell of carbon on the part.

The temperature used for pack carburizing should be above the upper transformation temperature line, so that the material structure changes to austenite. See Fig. 15-8. Therefore, the temperature employed might vary from 1500° − 1800°F (816° − 983°C), depending on the type of steel being treated. A temperature of 1700°F (927°C) would be a typical value.

During the first few hours, the case will grow at the rate of .010 in. − .030 in. (0.25 − 0.5 mm) per hr. This rate will drop down to .005 in. (0.13 mm) per hr. after five or six hours. After 10 hours, penetration is nearly exhausted. Thus, in an eight hour day, depths of .060 in. (1.52 mm) would be typical depending on the temperature and type of steel employed. See Fig. 15-9.

The pans must be capable of withstanding considerable abuse. Thermal abuse causes more boxes to "die" early than just plain physical abuse. Not only must the box withstand high temperature, but the repeated heating, cooling, reheating, recooling cycles cause early deterioration and internal stress.

Advantages of Pack Carburizing

Pack carburizing involves a minimal capital expense and is more foolproof than other surface-hardening techniques. It is especially practical when only a few small parts require case hardening at one time. See Fig. 15-7.

Disadvantages

Pack carburizing is a relatively slow and dirty case-hardening process.

GAS CARBURIZING

In GAS CARBURIZING, the parts are heated inside an oven or furnace, Fig. 15-10. The oven is filled with a carbonaceous gas. The gas may be natural gas, ethane, propane, butane, carbon monoxide, or a vaporized fluid hydrocarbon. Carbon atoms from the gas are enticed to attach onto the steel parts. Thus, the outer skin of the steel parts becomes carbon-filled. The longer the parts are left in this gas-filled oven, the deeper and more dense the carbon layer becomes.

After the surface of the parts has received enough carbon, the parts may be quenched to ob-

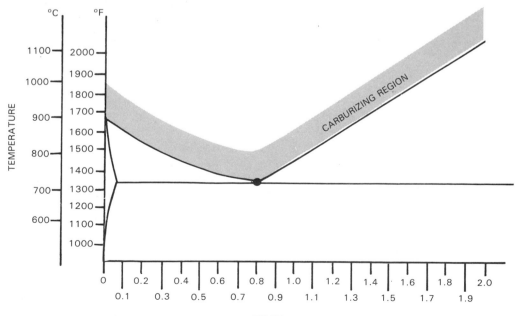

Fig. 15-8. Carburizing temperatures are slightly above upper transformation temperature values.

TYPICAL CASE GROWTH

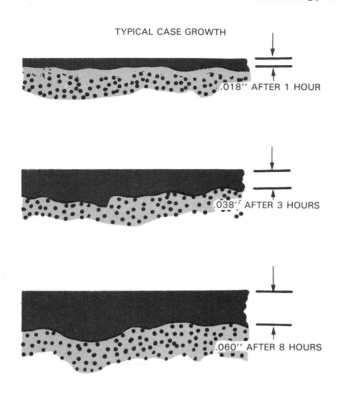

.018" AFTER 1 HOUR

.038" AFTER 3 HOURS

.060" AFTER 8 HOURS

Fig. 15-9. Three views show typical case-hardening growth during pack carburizing.

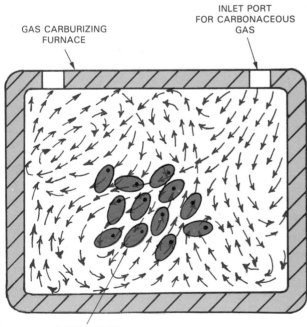

INLET PORT
FOR CARBONACEOUS
GAS

GAS CARBURIZING
FURNACE

PARTS TO BE
CASE HARDENED

Fig. 15-10. In gas carburizing, parts are heated inside an oven or furnace.

tain the hard case immediately. If machining is required, reheating and quenching can be delayed until after the machining is done.

As in pack carburizing, the parts that are gas carburized must be heated to the austenitic range in order to get the full carbon penetration. This temperature also will vary from 1500°F to 1800°F, depending on the type of low-carbon steel.

Fig. 15-11 shows a furnace that is used in gas carburizing. Both rotary and pit type furnaces have been used. In the rotary furnace, the parts are tumbled by rotating a retort. Today, continuous automatic furnaces are gaining in popularity because of efficiency and cost saving. Most larger furnaces use a fan or other circulating device to keep the gas constantly moving against the parts, Fig. 15-10. This encourages a more uniform carbon coat to be generated.

Advantages of Gas Carburizing Over Pack Carburizing

Gas carburizing is faster than pack carburizing. It requires less labor and handling. Also, when case hardening by the gas carburizing method, the case depth can be controlled more accurately. Finally, gas carburizing is more practical than pack carburizing for case hardening large quantities of parts.

Disadvantages

Both equipment and materials used in gas carburizing are more expensive than those employed in pack carburizing.

LIQUID CARBURIZING

The idea of LIQUID CARBURIZING is similar to the idea of pack and gas carburizing. Steel parts are put into contact with a carbon-rich material which permits carbon to deposit into their outside surface. However, with liquid carburizing, this material is not solid nor a gas. It is a liquid.

Fig. 15-11. Continuous carburizing furnace. (Sunbeam Equipment Corporation)

The carbon-rich liquid generally is a salt bath, Fig. 15-12. Sodium Cyanide (NaCN), Barium Cyanide ($BaCN_2$), Calcium Cyanide ($CaCN_2$), or other salts may be used.

Usually, the salt bath is heated by electricity. In most tanks, a stirring method is used to keep the solution moving uniformly over all of the parts.

Advantages of Liquid Carburizing Over Pack and Gas Carburizing

1. Liquid tends to transfer heat faster, so carbon is added rapidly during the first hour. Therefore, it is an efficient process for shallow cases.
2. The case is more uniform along its surface because liquids tend to flow more evenly along a surface than do gases or solids.
3. Liquid covers the immersed parts like a blanket, thus reducing the amount of oxidation contamination.

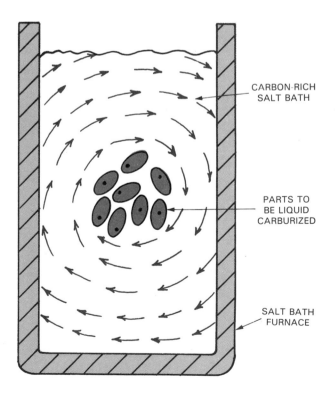

CARBON-RICH SALT BATH

PARTS TO BE LIQUID CARBURIZED

SALT BATH FURNACE

Fig. 15-12. In liquid carburizing, parts are immersed in a salt bath.

Disadvantages

1. With liquid carburizing, some nitrogen is absorbed from the cyanide salt along with the carbon. Nitrogen causes immediate hardening. Therefore, liquid carburized parts generally are not machined after carburizing.
2. Cyanide salts are poisonous. More than average care must be taken for safety of the operators.
3. Parts must be rinsed after liquid carburizing to prevent rusting. This takes time and requires more equipment.
4. Salt baths generally are relatively small chambers or tanks, so it is not convenient to immerse many large, odd-shaped parts into a liquid bath. Therefore, liquid carburizing usually is restricted to the case hardening of small parts.

NITRIDING

NITRIDING fills the outer skin of a steel part with nitrogen instead of carbon. It is a gaseous process, rather than a liquid or solid process. The nitrogenous gas usually used is ammonia (NH_3). When the nitrogen unites with the iron surface, several types of iron nitride are formed. Iron nitride is superhard, Fig. 15-13.

Nitriding has seven advantages and five disadvantages that make it unique. Therefore, nitriding may be the "best" or "worst" surface-hardening process to use, depending on the application.

Advantages of Nitriding

1. Hardness — Nitrogen offers the hardest case of all surface-hardening processes. Its case is more wear resistant than any other case. Hardnesses over 70 R_c have been attained.
2. Case-hardening Temperature — Nitrogen atoms will join the iron surface below the transformation temperature of steel. Temperatures of $900° - 1000°F$ ($482° - 538°C$) are normally used. This is the only surface-hardening process wherein this low temperature can produce a hardened case.

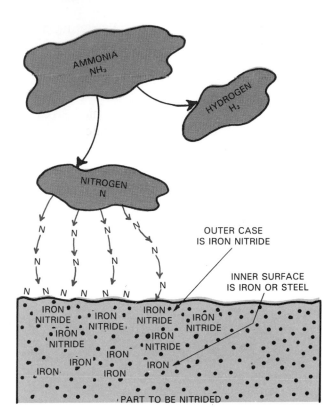

Fig. 15-13. In nitriding, parts to be nitrided come in contact with a nitrogenous gas which is usually ammonia.

Hardening at this low temperature offers more convenience, less processing cost, and less tendency toward distortion and cracking.

3. Immediate Hardness-No Heat Treat — After nitrogen joins the surface, the outer case is hard. No reheating and quenching is required, as is the case with carburizing. This saves time. It reduces distortion. However, it also makes post machining difficult.
4. Distortion — Less distortion, less warpage, less internal stress, less danger of cracking and better dimensional stability result. There are two reasons for this: lower case-hardening temperature (Advantage #2) and no quenching (Advantage #3). This is of special importance for complex parts with uneven sections that tend to distort easily.
5. Corrosion Resistance — Nitrided surfaces are more resistant to corrosion than other case-hardened parts. Humidity, salty atmospheres, water, oil, gasoline, and other corrosive agents cause less harmful effects.

6. Temperature Resistance — Reheating parts up to 1000° − 1100°F (538° − 593°C) for a short time does not affect a nitrided case. These temperatures would soften carburized cases. Prolonged heating at 600° − 800°F (316° − 427°C) will not affect a nitrided case, but will affect a carburized one. Thus, at elevated temperature, nitrided cases are more dimensionally stable.
7. Cleaning — Freshly nitrided parts do not require cleaning. Carburized parts do, to prevent corrosion.

Disadvantages

1. Slowness — Nitriding is the slowest of the case-hardening processes. You may talk in terms of days instead of hours. For this reason, nitriding is generally restricted to very thin cases, unless high hardness is extremely important. A .010 in. case might take 10 hours. A .030 in. case would take several days.
2. Cost — Nitriding is expensive. Ammonia gas is many times more expensive than the popular carburizing gases. The equipment, too, is expensive. Inexpensive low carbon steel cannot be used. The steel must contain alloys that react with the nitrogen to form the nitride compounds. These alloy steels cost more.
3. Size Growth — As the parts take on nitrogen, they begin to swell. This can affect dimensional accuracy. The amount of swelling can be estimated fairly closely. Often the part can be made undersized before case-hardening, to allow for this small change.
4. Machining — Little machining can be done to a part after it has been nitrided, because the hardening takes place immediately. Only a small amount of grinding is practical for a part in this hardened condition. You will recall that this was an advantage of pack and gas carburizing. Some machining could be postponed until after the first step of surface hardening was completed. This reduced the effects of distortion and warpage. Fortunately, the nitriding process itself induces very little distortion (Advantage #4).
5. Foolproofness — Close control is necessary with nitriding. The ammonia gas must be kept sealed in the heating chamber. An experienced operator is necessary to regulate the percent ammonia.

CARBO-NITRIDING

CARBO-NITRIDING is a combination of gas carburizing and nitriding. Both ammonia and natural gas are put into contact with the parts. The hard case ends up as a combination of iron carbide (from the carbon) and iron nitrides (from the nitrogen).

Advantages of Carbo-nitriding

Carbo-nitriding has an advantage over gas carburizing. Because of the presence of nitrogen, the austenite structure is altered. This change permits a lower transformation temperature and a slower cooling rate.

Carbo-nitriding temperatures then are about 1400° − 1700°F (760° − 927°C), about 100°F (55.6°C) lower than carburizing. See Fig. 15-14. Oil quenching can be used instead of water quenching because of the slower cooling rate that is permitted. Internal stresses are reduced, distortion and warpage are reduced, and there is less chance of cracks.

Case Depth with Carbo-nitriding

The rate of carburization by carbo-nitriding is slower than with carburizing. Therefore, carbo-nitriding is generally used for thin cases of about .005 in. − .010 in.

CYANIDING (LIQUID CARBO-NITRIDING)

CYANIDING is a popular case-hardening process. Its special features can be summarized in three statements:
1. Cyaniding is a liquid process that imparts both carbon and nitrogen to the surface.
2. Its fumes are deadly.

Fig. 15-14. Electrically heated carbo-nitriding furnace. (The Electric Furnace Company)

3. It is used primarily for obtaining fast, hard, thin coats.

Cyaniding is similar to liquid carburizing in that a molten salt bath is used. See Fig. 15-15. It is similar to carbo-nitriding in that both carbon and nitrogen are added to the surface. The salt bath is made up of a molten cyanide salt, usually sodium cyanide. Calcium cyanide or potassium cyanide are also used. Sodium cyanide melts at 1140°F (615°C), so it is in the liquid state when used.

Cyaniding is used primarily for very thin cases. During the first 30 min., the carbon and nitrogen penetrate the surface rapidly to a depth of about .005 in., depending on the concentration of the salt bath and the type of material. After the first 30 min., penetration slows up considerably. Therefore, cyaniding is seldom used for cases deeper than .010 in.

Since nitrogen is also added to the surface, very hard cases can be attained. Hardnesses of 65 Rockwell C are not uncommon.

Physically, cyaniding is often done using three tanks. The first tank contains a liquid and preheats the parts. The second tank contains a 30 percent sodium cyanide salt solution. Immersion in this tank provides the case hardening itself. The third tank is for quenching. Normally, parts that have been cyanided are quenched immediately. Since part of the hardening effect is due to nitrogen content, the quench does not have to be rapid. Therefore, oil quenching can generally be used instead of a water quenching. This eliminates some of the hazards of distortion and cracking.

Advantages of Cyaniding

Cyaniding is a fairly inexpensive case-hardening process because plain carbon steel can be used. Cyaniding is used where the application calls for a fast and thin case but a hard case.

Disadvantages

The cyaniding process can be very hazardous and should not be left in the hands of irrespon-

232

Fig. 15-15. This gas-fired furnace is used for cyaniding, liquid carburizing, or any application wherein a hot liquid salt is required. (Charles A. Hones Inc.)

sible employees. Cyanide salts are poisonous and their fumes can be fatal if inhaled. If the salt bath is permitted to come in contact with an open cut or wound, the results can be serious. Obviously the case-hardening area must be well vented when using a cyanide salt bath.

FLAME HARDENING

Flame hardening is different from any of the previous case-hardening methods that we have discussed. No carbon is added to the surface. No nitrogen is added to the surface. Only heat is added.

A direct flame from an oxyacetylene torch is brought into contact with the surface that is to be hardened. As soon as the surface is hot, quenching follows. The heating and quenching must

be done very rapidly. If the heat is given time to penetrate deeply into the part, there will no longer be just case hardening, but the core will be hardened as well.

If enough heat is applied to the surface to reach the upper transformation temperature before quenching, then martensite will be formed and you will have a hard outer shell.

Obviously, this process is extremely well suited for certain applications but restricted from other types of applications. There are certain advantages and disadvantages that control the suitability or unsuitability of flame hardening.

Advantages of Flame Hardening

1. Deep Cases — Flame hardening is a very rapid and efficient method for providing deep cases up to about 1/4 in.
2. Zone Hardening — Zone hardening is possible with this method. When only a small portion of a part is to be hardened, Fig. 15-16, it is not necessary to harden the entire part. This is a disadvantage of all of the previous methods discussed. With flame hardening, heat can be applied merely to the small critical "zone" and quenched. This reduces the tendency of distortion since only a small portion of the part is heated. Tempering is still recommended after quenching in order to reduce the localized stresses, but distortion of the entire part is kept to a minimum.
3. Small Quantities — The flame-hardening method lends itself quite readily to small quantities of parts. It usually is not economically feasible to heat up a large oven or salt bath when only a few parts need hardening.
4. Cost of Capital Equipment — The cost of capital equipment is negligible, assuming that most companies have a welding torch available. No expensive ovens are required for flame hardening. No expensive chemicals or gases are required.
5. Large Parts — This method lends itself to the hardening of large bulky parts that will

Fig. 15-16. Flame hardening is used to zone harden these wrenches. Jaw area of these four wrenches receives flame-hardening treatment while rest of wrench remains relatively soft and impact resistant. Wrenches are stationary as flame is applied, then quenched afterward with oil.
(Tocco Division, Park-Ohio Industries, Inc.)

not fit into a furnace or liquid case-hardening tank. Also, large, heavy parts that cannot be transported across the factory conveniently can be flame hardened on location because the torch can be taken "to" them.

Disadvantages

1. Thin Cases — Because the heat penetration into the metal is very difficult to control accurately, flame hardening usually is not used for thin cases.
2. Types of Steel — Only certain steels can be case hardened by this method. Since no additional carbon or nitrogen is added to the surface, the source of the hardening must come from the metal itself. The steel must contain more than the average amount of carbon. Low carbon steel cannot be hardened by this method. Medium carbon steels with 0.35 to 0.60 percent carbon are most commonly used. These steels are slightly more expensive than low carbon steels that can be case hardened by the other methods.

Hardness Level

The maximum hardness that usually is obtained by flame hardening is less than that of other case-hardening methods, since you are relying entirely on carbon content that is already in the material. Hardnesses of 50-60 R_c can be obtained.

Automation

The amount of automation involved in flame hardening varies from none to full automation. Some parts are flame hardened with the use of simply a welding torch and a water hose. At the other extreme, highly mechanized, automated systems have been designed to flame-harden parts on a mass production basis.

A semiautomated process that is quite popular involves a tool that is a combination torch and water spray gun. With this method, a fairly uniform depth is attained between parts at reasonable expense. Figs. 15-17 and 15-18 illustrate an industrial application of this process wherein gear teeth are flame hardened.

Occasionally, instead of using a quenching spray after the heating process, the parts may be immersed in a quenching medium.

Applications

Common applications of flame hardening include any application wherein the depth of hardness is not critical or where only a small zone of the part requires hardening. Examples are gear teeth, cylindrical pins, lathe beds, cam surfaces, engine push rods, pulleys, and sprocket teeth.

INDUCTION HARDENING

Induction hardening could be called "high class flame hardening." No carbon is added. No nitrogen is added. Only heat is added. Like flame hardening, only the outer surface is heated above the upper transformation temperature before the part is quenched.

Fig. 15-17. Flame hardening of gear teeth is automated. Gear in center rotates past stationary flaming heads around its outer periphery which heat the teeth. Flames shown here are coming from pilot jets before flame-hardening operation is started. (Tocco Division, Park-Ohio Industries, Inc.)

Fig. 15-18. This view shows flame hardening of gear teeth in action. As gear is rotating, numerous flame jets (both in foreground and background) can be seen firing jets of flame and applying heat to teeth. Two pilot jets can still be seen at left. Following heating of each gear tooth, a quench spray would be applied to gear. (Tocco Division, Park-Ohio Industries, Inc.)

The big difference between flame hardening and induction hardening, is the fascinating way in which the heating is accomplished. A high fre-

quency electrical current flows through a coil of wire, Fig. 15-19, causing a magnetic field around the steel part that is to be hardened. This causes eddy currents to pass through the metal. Due to a phenomenon called "skin effect," the outer surface of the steel part becomes hot.

The part to be hardened is surrounded by the electrical coil which acts like the primary coil of a transformer. See Figs. 15-20, 15-21, 15-22. As high frequency electrical current, 3000 to 1,000,000 cycles per second (hertz), passes through the primary coil, a secondary current tries to flow through the parts. The electrical resistance of the part to this current flow causes heat. Due to this skin effect, the current and the heat stay only on the outer surface or the skin of the steel. After the skin of the steel part has received sufficient heat, the part is quenched in water or oil.

Fig. 15-19. Gears are induction hardened by surrounding gear with an electrical coil in a ring. (J.W. Rex Company)

Advantages of Induction Hardening

1. Induction hardening is fast. The surface of the part can be heated in one to five seconds. This makes it very practical for automation. The part cost in large quantities is very inexpensive.

2. No warm-up time is required prior to the meeting between the electric coil and the part.

3. Irregular shapes can be handled quite readily with induction hardening. The skin current can dip into crevices and holes, and it will harden interior surfaces.

4. The thickness of the case can be controlled more accurately with this method than with any other. The depth can be controlled by varying the frequency, the current, and the time that the coil is in contact with the part. The higher the frequency, the more the current tends to flow over the outer skin only.

5. Induction hardening is generally used for thin cases. Depths of 1/8 in. can be reached by leaving the current in contact with the surface for a longer period of time and operating at lower frequencies.

6. Induction hardening offers outstanding resistance to warpage, distortion, oxidation, and scale. This is because of the short period of time involved in the heating process and the fact that only a small portion of the part needs to be heated at all.

7. Accuracy is possible with this process even if an unskilled operator runs the equipment. However, setting up the induction-hardening job requires a very experienced and knowledgeable technician.

8. The entire induction-hardening operation is cleaner and less messy than any of the other methods.

Disadvantages

1. Since no carbon or nitrogen is added to the surface of the part, a more expensive steel is required. A medium carbon steel with .35 to .60 percent carbon most often is used.

2. The hardness that is obtained is dependent on the carbon content of the steel that you start with. Therefore, hardnesses over 60 with induction hardening cannot be obtained unless expensive high alloy steel is used.

Applications

Practical examples of induction hardening as a case-hardening method include: irregular shaped parts such as cams, gear teeth, and shaft splines; bearing surfaces of automotive crankshafts; pump shafts; piston rods; ball and roller bearings; chain links; firearm parts.

Other practical examples are shown in Fig. 15-23 through 15-34.

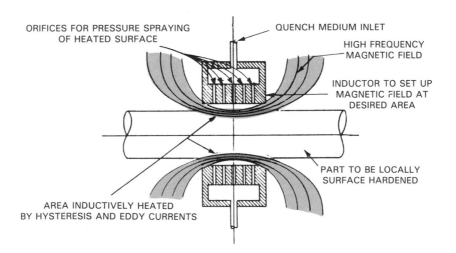

Fig. 15-20. Induction hardening. This shows a cross-sectional view of an inductor and a workpiece of bar stock. Magnetic lines of force emitted from inductor cut through surface of workpiece. Dark areas depict area heated by magnetic lines of force. Long narrow holes in inductor serve as quench orifices so that a quench can be injected after part is heated. Dark areas shown will become hard after induction hardening is completed. (Tocco Division, Park-Ohio Industries, Inc.)

Fig. 15-21. This coil contains both an inductor coil for induction heating and four orifices to provide quenching all in one package. Part to be induction hardened passes through center of coil where approximately 1000 amps of current flow through inductor. Following heating by induction current, a liquid quench flows through four rows of inlets around inner periphery of coil. Thus, this one package contains both inductor coil for heating and inlets for quenching.
(Tocco Division, Park-Ohio Industries, Inc.)

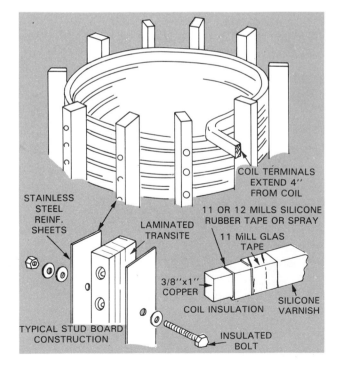

Fig. 15-22. This drawing shows configuration and construction of a typical induction heating coil that is used for induction hardening. (Ajax Magnethermic Corp.)

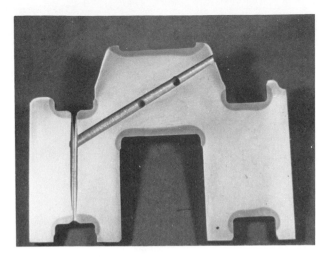

Fig. 15-23. This cross section of a crankshaft shows wherein induction hardening was applied to two main bearing surfaces and one crankpin bearing surface. Corner fillets are also hardened. Hardened areas are darker in color.
(Tocco Division, Park-Ohio Industries, Inc.)

Fig. 15-24. This is a cross section of a front wheel spindle that has been induction hardened. This spindle is used on one of newer compact automobiles.
(Tocco Division, Park-Ohio Industries, Inc.)

SELECTION OF CASE-HARDENING METHOD

Each of these eight case-hardening processes has advantages and disadvantages over the others. The decision of which one to use will depend on the quantities of parts involved, the accuracy and depth of case required, the shape and size of the parts, and the equipment and facilities available to do the case hardening.

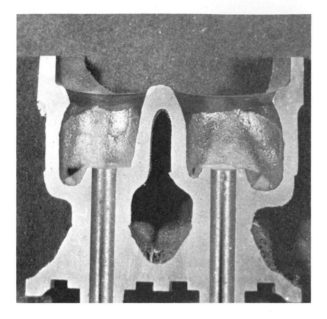

Fig. 15-25. This is a cross section of a cylinder head that was hardened by induction hardening. Two lightly colored areas are valve seats which are hardened areas.
(Tocco Division, Park-Ohio Industries, Inc.)

Fig. 15-26. This familiar ballpeen hammer has been induction hardened only in two areas where severe impact is encountered. Dark areas represent two areas that have been hardened. Lighter area in middle still retains softer ductile qualities to resist shock. (Tocco Division, Park-Ohio Industries, Inc.)

Fig. 15-27. Induction hardening is used to harden both this internal and external hub. This part is used on bulldozers. (Ajax Magnethermic Corporation)

Fig. 15-28. These sprocket teeth are hardened one tooth at a time. These sprockets are used for off-highway equipment. (Ajax Magnethermic Corporation)

Fig. 15-29. In this internal gear, one tooth is hardened at a time by induction-hardening process. (Ajax Magnethermic Corporation)

Fig. 15-30. Induction heating is used in this pipe-weld heat-treating application.
(Ajax Magnethermic Corporation)

Fig. 15-31. These mill rolls, up to 30 ton in weight, are case
hardened by induction-hardening method.
(Ajax Magnethermic Corporation)

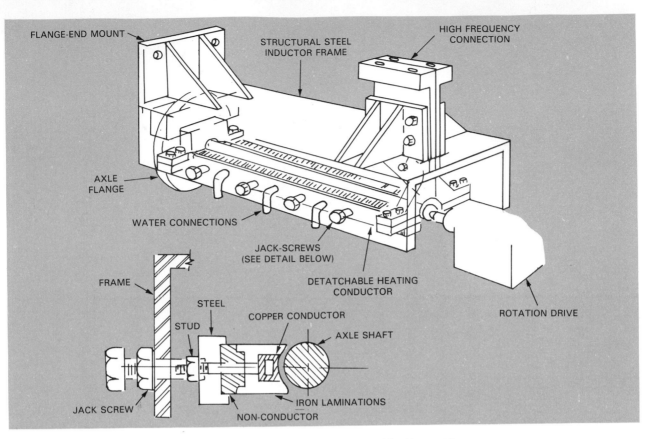

Fig. 15-32. Axles are induction hardened in this machine.
(Ajax Magnethermic Corp.)

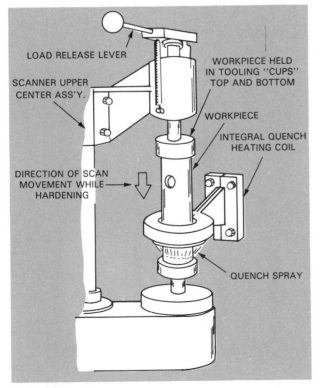

Fig. 15-33. Pins are case hardened in this machine by an integral induction-heating coil and quench spray combination. Coil-spray unit moves vertically over surface of pin to produce uniform induction hardening. (Ajax Magnethermic Corp.)

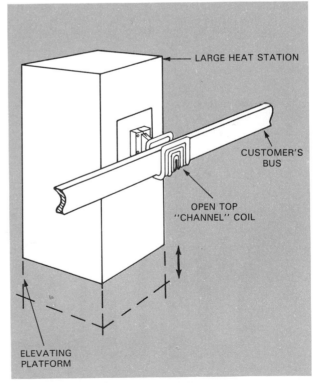

Fig. 15-34. Induction heating coils have many uses in addition to induction hardening. This heat station uses an induction coil to heat a "bus" hot enough for brazing.
(Ajax Magnethermic Corp.)

CASE-HARDENING METHOD	What Is Added To Surface	Order Of Hardness Attainable	Speed Of Addition Of Case	Temperature	Hazardness	Equipment And Tooling Cost	Piece-Part Cost In High Quantities	Piece-Part Cost In Low Quantities	Is Quenching Necessary After First Step?	Danger Of Distortion	Can Machining Be Done After First Step?	Control And Accuracy	Depth (in.) Per Unit Time (A Typical Case)
PACK CARBURIZING	Carbon	4 (tie)	Medium fast for small quantities	Highest (Austenite temperature required)	Little hazard	Very Low	High	Very Low	Yes	Some	Yes	Poor	.010 per first hour
GAS CARBURIZING	Carbon	4 (tie)	Fairly fast in high quantities	Highest (Austenite temperature required)	Fairly poisonous gas	High (furnace required)	Low	High	Yes	Some	Yes	Good	.013 per first hour
LIQUID CARBURIZING	Carbon	4 (tie)	Fast in medium sized quantities	Highest (Austenite temperature required)	Fairly poisonous	Medium	Medium	Medium	Yes	Some	Some	Good	.018 per first hour
NITRIDING	Nitrogen	1 Hardest	Very slow	Lowest 900-1000 °F	Medium caution required	High (furnace required)	Medium	High	No	Least	Difficult	Very Good	.010 per first 10 hours
CARBO-NITRIDING	Carbon and Nitrogen	2 (tie)	Medium fast in high quantities	2nd lowest (tie)	Fairly poisonous gas	High (furnace required)	Low	High	Yes	Little	Difficult	Good	.008 per first hour
CYANIDING	Carbon and Nitrogen	2 (tie)	Fast in medium sized quantities	2nd lowest (tie)	Very poisonous fumes	Medium	Medium	Low	Yes	Little	Difficult	Good	.010 per first hour
FLAME HARDENING	Only Heat	8 (hardness depends on material)	Fast in small quantities	Highest (Austenite temperature required)	Some caution needed with flame	Low	High unless automated	Low	Yes	Yes	Difficult	Poor	Instantly for thick cases
INDUCTION HARDENING	Only Heat	7 (hardness depends on material)	Fast in high quantities	Highest (Austenite temperature required)	Caution required for expensive equipment	High (electric equipment required)	Lowest	Varies	Yes	Little	Difficult	Best	Instantly for thin cases

Fig. 15-35. Chart details characteristics, advantages, and disadvantages of eight common methods of case hardening.

The table in Fig. 15-35 summarizes the comparative advantages and disadvantages of each method. Bear in mind that the table refers to the average typical case. With slight alterations, a case-hardening method can be applied beyond its typical applications.

TEST YOUR KNOWLEDGE

Write your answers on a separate sheet of paper. Do not write in this book.

Use the following words to answer the 13 questions below. Some of these words may be used more than once, some of the words may not be used at all.

carbo-nitriding	induction hardening
cyaniding	liquid carburizing
flame hardening	nitriding
gas carburizing	pack carburizing

1. The case-hardening process that is capable of producing the hardest case is known as

   _____ .

2. The case-hardening method that involves both carbon and nitrogen, and is a liquid method of hardening, is known as _____ .

3. The two hardening methods that do not involve the addition of either carbon or nitrogen are known as _____ and

   _____ .

4. The case-hardening method wherein the material need not be heated over 1000 deg. is known as _____ .

5. The case-hardening method which is the most foolproof, least hazardous, and involves the least amount of skill of the operator is called _____ .

6. Of the eight common methods of case hardening, how many of them impart nitrogen to the surface of the specimen? (Answer should be a number, not a list of methods.)

7. The case-hardening process that may include the use of bone, leather, or charcoal is known as _____ .

8. The case-hardening process that is normally the "slowest" is known as _____ .

9. A case-hardening method that can be employed when the parts are extremely large, and are located a long way from the manufacturing area of the building itself, is known as _____ .

10. Two case-hardening methods wherein the equipment required is comparatively less expensive are _____ and _____ .

11. The case-hardening method that is hazardous because of poisonous fumes is _____ .

12. The case-hardening method that involves the use of a torch is known as _____ .

13. The case-hardening method that may involve the use of electricity is known as _____ .

DICTIONARY OF TERMS

A

AGITATION: Rapid vibration of metal parts during quenching, in order to attain a high degree of hardness.

ALLOY: 1. A material which is dissolved in another metal in a solid solution. 2. Material that results when two or more elements combine in a solid solution.

ALLOY STEEL: A steel that contains more than the average amount of alloys. Alloys put into these steels give the steel special properties.

ANNEALING: Slow cooling of metal by controlled cooling within a furnace or oven. Furnace or oven temperature is reduced very slowly in order to attain a high degree of softness in the metal.

ATOM: Smallest possible part of an element that still has the characteristics of that element.

AUSTEMPERING: A special tempering technique wherein the tempering is not as drastic. Dwell temperature is maintained for a long period of time slightly above the martensitic transformation temperature.

AUSTENITE: One of the basic steel structures wherein carbon is dissolved in iron. Austenite occurs at elevated temperatures.

AUSTENITIC: Steel which has the structural form of austenite.

B

BAINITE: A steel structure which is harder than pearlite, cementite, or ferrite and more ductile than martensite. In a sense, it is a happy medium between martensite and softer steel structures.

BASIC OXYGEN FURNACE: A basic steel-making furnace that utilizes an oxygen blast at supersonic speeds to intensify the heat. Today, the basic oxygen furnace (BOF) has become the most widely used process in the manufacture of steel.

BESSEMER CONVERTER: One of the basic steel-making furnaces. This process utilizes a furnace in which molten pig iron is refined by a burning gas. At one time, the Bessemer Converter was used extensively in the manufacture of steel; however, today it is used minimally.

BHN: Unit of hardness used in the Brinell hardness testing method. BHN stands for Brinell hardness number.

BILLET: A semi-completed metal form made from a bloom and smaller in cross-sectional size than a bloom.

BLAST FURNACE: A large furnace used to convert iron ore into pig iron.

BLOOM: A semi-completed metal form in which the cross-section is relatively square.

BODY CENTERED CUBIC: One of the common types of unit cells. This arrangement is typical of the ferritic form of iron.

BODY CENTERED TETRAGONAL: One of the common types of unit cells. This arrangement is typical of the martensitic form of iron.

BOF: Same as BASIC OXYGEN FURNACE.

BRINE: A solution of salt in water which is occasionally used as a quenching medium in the cooling of steel.

BRINELL: A common hardness testing method.

BRITTLENESS: Tendency to stretch or deform very little before fracture.

C

C CURVES: Transformation curves utilized on isothermal transformation diagrams.

CARBON STEEL: Steel that contains comparatively less alloys than other steels.

CARBO-NITRIDING: A surface-hardening technique wherein the surface is impregnated with both carbon and nitrogen. This is a gaseous process.

CARBURIZING: A surface-hardening technique wherein the surface is impregnated with carbon.

CASE HARDENING: Same as SURFACE HARDENING.

CAST IRON: A material containing primarily iron, 2.0 to 6.0 percent carbon, and often small amounts of silicon and other elements.

CEMENTITE: One of the basic steel structures wherein carbon is dissolved in iron. Cementite occurs at room temperature when the steel has either (a) not previously been heat treated, or (b) has been cooled slowly after being heated and transformed to austenite. Cementite contains more than 0.8 percent carbon.

CHECKER CHAMBER: Brick structure on either side of an open hearth furnace used to retain the heat of the exhaust gases.

CLEAVAGE FAILURE: A brittle failure wherein little stretch of the metal occurs.

CLOSE PACKED HEXAGONAL: One of the common types of unit cells. Iron does not take this crystalline form.

COEFFICIENT OF THERMAL EXPANSION: Characteristic of a material which describes the amount of expansion that a material undergoes due to heat.

COKE: Purified coal used in the manufacture of iron and steel.

COMPOUND: A material that is composed of two or more elements that are chemically joined.

COMPRESSIVE STRENGTH: A material's ability to withstand a "pressing" or "squeezing together" type of stress.

CONTINUOUS CASTING: A modern method of steel manufacture wherein steel is continuously poured into either shapes, blooms, slabs, or billets.

CORROSION RESISTANCE: Resistance of a material to protect itself against chemical attack by the environment.

CRYSTAL: Same as GRAIN, DENDRITE, or UNIT CELL. Usually, crystal is used as a synonym for grain.

CRYSTALLIZATION: Formation of crystals as a metal solidifies. The atoms assume definite positions in the crystal lattice.

CUPOLA: A furnace commonly used in the manufacture of cast iron. It uses coal as its fuel.

CYANIDING: A surface-hardening technique wherein the surface is impregnated with both carbon and nitrogen. This is a liquid process.

D

DEFORMATION: Amount that a material increases or decreases in length as it is loaded.

DENDRITE: A growing colony of metal which has not completed its lattice growth. A dendrite is larger than a unit cell, but smaller than a grain.

DIELECTRICAL STRENGTH: Electrical property of a material which is a measure of its resistance to breakdown from a large voltage over a prolonged time period.

DPH: Unit of hardness used in the Vicker's hardness testing method. DPH stands for diamond pyramidal hardness.

DRAWING: Same as TEMPERING.

DUCTILE CAST IRON: One of the five categories of cast iron. It has superior properties to gray and white cast iron. It is also called "nodular cast iron."

DUCTILITY: Tendency to stretch or deform appreciably before fracture.

E

ELECTRIC ARC FURNACE: One of the basic steel-making furnaces. It is especially valuable in the production of high alloy steel.

ELECTRIC INDUCTION FURNACE: A furnace used in the manufacture of cast iron. It utilizes electricity as its source of power.

ELECTRICAL CONDUCTIVITY: Ability of a material to permit electricity to flow through it. This is the opposite of electrical resistance.

ELECTRICAL RESISTANCE: Ability of a material to resist the flow of electricity. This is the opposite of electrical conductivity.

ELEMENT: A simple, pure substance that is made up of one kind of material.

ENDURANCE STRENGTH: Ability of a material to withstand a repeated stress loading.

ETCHING: Application of acid onto a smoothly polished metal surface, to alter the crystalline depth of different metals, so that the metal structure can be observed under a microscope.

EUTECTOID POINT: Point on an iron carbon diagram wherein the upper transformation temperature line, the lower transformation temperature line, and 0.8 percent carbon pearlitic line, all intersect.

F

FACE CENTERED CUBIC: One of the common types of unit cells. This arrangement is typical of the austenitic form of iron.

FATIGUE STRENGTH: Ability of a material to withstand a repeated loading.

FERRITE: One of the basic steel structures wherein carbon is dissolved in iron. Ferrite occurs at room temperature when the steel has either (a) not previously been heat treated, or (b) has been cooled slowly after being heated and transformed to austenite. Ferrite steel contains a very small percentage of carbon.

FERRITIC: Steel which has the structural form of ferrite.

FILE HARDNESS TEST: A fast, simple hardness test in which the edge of a file is scraped across a test material to see if it will scratch the surface.

FLAME HARDENING: A surface-hardening technique wherein the hardness is attained by heating the outside surface with a direct flame and then cooling it by a sudden application of a cold liquid.

FLEXURE STRENGTH: Bending strength. It generally involves tensile stress on one side of the material and compressive stress on the other.

FULL ANNEALING: Annealing initiated at a temperature above the upper transformation temperature. It provides the slowest cooling of any heat-treating process.

G

GAS CARBURIZING: A carburizing surface-hardening technique wherein the carbon that is impregnated into the surface of the metal comes from a gaseous source.

GRAIN: Any portion of a solid which has external boundaries and an internal atomic lattice arrangement that is regular.

GRAY CAST IRON: One of the five types of cast iron. It is the most widely used, is one of the least expensive, and has comparatively less tensile strength.

H

HARDNESS: A measure of the resistance to deformation or penetration.

HEAT TREATING: Heating and cooling metal to prescribed temperature limits for the purpose of changing the properties or behavior of the metal.

HEMATITE: One of the common types of iron ore.

HIGH CARBON STEEL: Carbon steel that contains between approximately 0.5 percent and 2.0 percent carbon.

HOT METAL: Semi-refined, molten iron which is a product of the blast furnace. It is kept in the molten state and then transferred directly to a steel-making furnace. This is also referred to as "liquid pig iron."

HYPEREUTECTOID REGION: Any steel on an iron carbon diagram that falls to the right of the 0.8 percent carbon line.

HYPOEUTECTOID REGION: Any steel on an iron carbon diagram that falls to the left of the 0.8 percent carbon line.

I

IMPACT STRENGTH: Ability of a material to resist shock. It depends on both strength of the material and ductility of the material.

INDUCTION HARDENING: A surface-hardening technique wherein the hardness of the surface is attained by heating it with an electromagnetic field and then cooling it by a sudden application of a cold liquid.

INGOT: The initial cast shape of steel before it is rolled into a workable shape.

IRON-CARBON PHASE DIAGRAM: A graphic representation of the transformation temperatures and carbon composition of various steel alloys. This graph plots temperature VS carbon content of steel or iron. From this diagram, the observer can determine the structure of a given steel.

IRON ORE: A mineral which is mined from the ground and has a high iron content. It is used as a basic ingredient in the manufacture of all alloys of iron and steel.

ISOTHERMAL QUENCHING AND TEMPERING: A special tempering technique wherein the metal is tempered before quenching has completely taken place.

ISOTHERMAL TRANSFORMATION DIAGRAM: A graph of temperature VS time for the process of cooling metal. From this type of diagram, the final structure of a metal can be predicted.

I-T DIAGRAM: Same as ISOTHERMAL TRANSFORMATION DIAGRAM.

K

KNOOP: A common hardness testing method.

L

LIMESTONE: A mineral that is mined from the ground and is used as a basic ingredient in the manufacture of iron and steel. Its basic purpose is to remove impurities.

LIQUID CARBURIZING: A carburizing surface-hardening technique wherein the carbon that is impregnated into the surface of the metal comes from a liquid source.

LOW ALLOY STRUCTURAL STEEL: An alloy steel used for structural work which contains less alloys than most alloy steels. It contains more alloys and is more expensive than carbon steel.

LOW CARBON STEEL: Carbon steel that contains between approximately .05 percent and 0.35 percent carbon.

LOWER TRANSFORMATION TEMPERATURE: Temperature at which a metal structure starts changing to a different structure as it is being heated from a lower temperature. When the metal is cooled from an elevated temperature, above the lower transformation temperature, it is that temperature at which structural changes are theoretically completed.

M

MAGNATITE: One of the types of iron ore.

MALLEABLE CAST IRON: One of the five categories of cast iron which has superior properties to gray and white cast iron.

MARAGING STEEL: A special steel that contains a high quantity of nickel and a small quantity of carbon. It has very high strength and toughness.

MARTEMPERING: A special tempering technique wherein the quenching action is slightly less drastic than regular tempering.

MARTENSITE: One of the basic steel structures wherein carbon is dissolved in iron. Martensite occurs at room temperature when the steel has previously been heated and transformed to austenite, and then rapidly quenched.

MARTENSITIC: Steel which has the structural form of martensite.

MEDIUM CARBON STEEL: Carbon steel that contains between approximately 0.35 percent and 0.5 percent carbon.

MELTING POINT: Temperature at which a material turns from a solid to a liquid.

METAL: An element that has several metallic properties such as ability to conduct electricity, hardness, heaviness, nontransparency, etc.

METALLURGY: Science that explains the properties, behavior, and internal structure of metals. Metallurgy also teaches us what to do to metals to get the best use out of them.

MIXTURE: A material composed of two or more elements or compounds, which are mixed together and not chemically joined.

MOH: An old hardness testing scale.

MOLECULE: Smallest part of a compound that still has the characteristics of that compound.

N

NITRIDING: A surface-hardening technique

wherein the surface is impregnated with nitrogen.

NODULAR CAST IRON: Same as DUCTILE CAST IRON.

NORMALIZING: Slow cooling of metal, by allowing the metal to cool freely at room temperature.

O

OPEN HEARTH FURNACE: A basic steel-making furnace which utilizes a giant hearth exposed to a powerful gas flame that melts the ingredients of the hearth. At one time, the open hearth furnace was the primary process in the manufacture of steel. Today, it is still used, but not as extensively as the basic oxygen furnace.

P

PACK CARBURIZING: A carburizing surface-hardening technique wherein the carbon that is impregnated into the surface of the metal comes from a solid source.

PEARLITE: One of the basic steel structures wherein carbon is dissolved in iron. Pearlite occurs at room temperature when the steel has either (a) not previously been heat treated, or (b) has been cooled slowly after being heated and transformed to austenite. Pearlite steel contains approximately 0.8 percent carbon.

PENETRATION HARDNESS: Hardness obtained by a testing method which employs very accurate techniques. A precision penetrator is utilized to indent the metal surface.

PENETRATOR: Ball or pointer that is used in a hardness testing machine to penetrate or indent the sample.

PERCENT ELONGATION: A common reference that indicates ductility of a material. It is percent of elongation per unit length.

PIG IRON: Semi-refined iron which is produced by the blast furnace. It is also referred to as "hot metal."

PROCESS ANNEALING: A special type of annealing wherein the metal is heated to a temperature below the lower transformation temperature and then cooled in a furnace or oven. This process is faster than full annealing, but it has less hardening effect on the metal.

Q

QUENCHING: Rapid cooling of metal, usually by plunging metal into a cold liquid bath.

QUENCH AND TEMPER STRUCTURAL STEELS: Structural steels which are stronger, more expensive, and have higher alloy content than low alloy structural steel.

QUENCHING MEDIUM: Cold bath, usually a liquid, into which metal is plunged during the quenching process.

R

REFRACTORY MATERIAL: Material capable of enduring high temperature.

REGIONS OF TRANSFORMATION: Four regions on an I-T diagram that indicate where transformations to the various structures of the steel can take place.

ROCKWELL: Most commonly used hardness testing method.

ROCKWELL SUPERFICIAL: A common hardness testing method.

ROLLING MILL: The part of a steel-making plant wherein a series of large, hard rollers compress steel ingots into different shapes.

S

SCLEROSCOPE: A common hardness testing machine called the "Shore Scleroscope."

SCRATCH HARDNESS: A fast and crude method of measuring hardness without a precision machine being used. The metal is normally scratched by the edge of a tool or object. Then the hardness of the sample is judged by its scratch resistance.

SHEAR FAILURE: Failure of a ductile material in which atoms slide past each other within the crystals. Examples are SLIP or TWINNING failure.

SHEAR STRENGTH: A material's ability to resist a "sliding past" type of stress.

SHORE: A common hardness testing method.

The hardness testing machine used is commonly called the "Shore Scleroscope."

SLAB: A semi-completed steel form in which the width is appreciably larger than the thickness.

SLAG: A product of the iron and steel-making furnaces. It is basically considered to be a waste product, although today there are a few practical uses for it.

SLIP: A shear type of failure of a ductile material wherein rows of atoms slide past each other as fracture occurs.

SOAKING PITS: Large ovens where ingots are heated for several hours before being transported to the rolling mill. Purpose of the soaking pit is to obtain more uniform properties in the metal.

SOLID SOLUTION: A solution in which both the solvent and solute are solid materials at room temperature.

SOLUTE: Substance which is dissolved in another substance in a solution.

SOLUTION: A special type of mixture wherein one of the substances is thoroughly dissolved in the other.

SOLVENT: Substance which dissolves or overpowers another material in a solution. It usually is a liquid.

SONODUR: A hardness testing method.

SPACE LATTICE: Regular pattern or arrangement of the atoms in a crystal.

SPECIAL ALLOY CAST IRON: One of the five categories of cast iron. It may contain high percentages of nickel, copper, chromium, or other alloys.

SPHEROIDIZING: A special type of annealing wherein the metal is heated to a temperature below the lower transformation temperature and then cooled in a furnace or oven. This differs from process annealing in that, higher carbon steels are involved and the microscopic structural pattern shows many unique tiny spheres.

SPRING STEEL: A special category of steel that has unusually good elasticity and strength.

STAINLESS STEEL: A type of alloy steel that has outstanding corrosion resistance. All stainless steels contain high quantities of chromium, and some contain high quantities of nickel as well.

STEEL: A material composed primarily of iron, less than 2.0 percent carbon and, normally, small percentages of other elements.

STRAIN HARDENING: A hardening phenomenon that takes place as metal is repetitively loaded.

STRAND CASTING: Continuous casting.

STRENGTH: Ability of a metal to resist forces or loads. Strength is normally measured in force per unit area.

STRESS: The force per unit area imparted to a material. Stress also refers to the available strength or ability to withstand a force per unit area.

STRESS RELIEVING: Same as tempering, but for the primary purpose of relieving internal stresses.

STRIPPING: Removing ingots from the ingot molds.

SURFACE HARDENING: A heat-treating process wherein the surface of the metal becomes very hard but the interior core remains soft and ductile.

T

TACONITE: A popular form of iron ore which is very plentiful but has a comparatively low iron content compared to other iron ores.

TEEMING: Pouring molten steel into ingot molds.

TEMPERATURE-TIME-LINE: Line on I-T diagram that follows the path of the temperature of the steel with respect to time after quenching begins.

TEMPERING: The reheating of metal after it has been quenched, for the purpose of slightly softening it and making it more stress-free, distortion-free, and crack-free.

TENSILE STRENGTH: A material's ability to withstand stress in tension, or pulling apart.

THERMAL CONDUCTIVITY: The ability of a material to transmit heat.

TIME-LINE: Same as TEMPERATURE-TIME-LINE.

TIME SCALE: Scale in seconds that appears at bottom of curve on I-T diagram.

TOOL STEEL: A group of alloy steels which

may have high strength, hardness, wear resistance, etc.

TOUGHNESS: Ability of a material to resist shock. Toughness depends on both strength and ductility of the material.

TRANSFER TEMPERATURE RANGE: Thermal region between the upper and lower transformation temperatures of a metal wherein structural transformation takes place.

T-T-T DIAGRAM: Same as ISOTHERMAL TRANSFORMATION DIAGRAM. T-T-T stands for temperature, time, and transformation.

TUNDISH: Top part of a continuous casting machine which holds the molten metal.

TWINNING: A shear type of failure of a ductile material wherein two lines of atoms simultaneously slide past each other, forming mirror lines and twinning planes.

U

UNIT CELL: The most fundamental arrangement of atoms within a space lattice.

UPPER TRANSFORMATION TEMPERA-TURE: Temperature at which a metal structure starts changing to a different structure as it is being cooled from an elevated temperature above the upper transformation temperature. When the metal is heated from a lower temperature, it is that temperature at which the structural changes are theoretically completed.

V

VICKERS: A common hardness testing method.

W

WEAR: Ability of a material to withstand wearing away by frictional scratching, scoring, galling, scuffing, seizing, pitting, fretting, etc.

WHITE CAST IRON: One of the five categories of cast iron. It is comparatively hard and brittle.

WORK HARDENING: A hardening phenomenon that takes place as metal is repetitively loaded.

WROUGHT IRON: A material that is composed almost entirely of iron. It contains essentially no carbon.

INDEX

Cementite,
 appearance, 155, 156
 structure, 140, 141, 155
 transformation, 145, 146, 151
Cementite-pearlite structure, 157
Checker chamber, 56
Chemical properties, 110, 116
Chemical terms, 11-19
Cleavage, 133, 134
Close packed hexagonal space lattice, 125
Coefficient of thermal expansion, 117, 119
Compound, 12, 13
Compressive strength, 112
Conductivity, electrical, 117
Conductivity, thermal, 117, 119
Continuous casting, 70-72
Corrosion resistance, 116
Cracking, 137, 138, 213, 220
Crushers, 43
Crystallization, 122
Crystals, 18, 122, 137, 138
Crystal structure, 122
 crystal growth, 128, 129
 crystallization, 122
 refinement, 183
 space lattice, 125
 transformation temperature, 127, 128
Cupola, 73-76
Curved mold method, 70
Cyaniding (liquid carbo-nitriding),
 231, 232, 233

D

Deformation, 132
Dendrites, 128
Deposits, iron ore, 43
Dielectric strength, 117
Distortion, 8, 9, 217, 220
DPH, 84, 90
 diamond pyramidal hardness, 90
 formula, 90
Drawing, 213
Ductile material, 132
 cast iron, 37, 38
 failures, 134-137
Ductility, 113, 114, 151

E

Electrical conductivity, 117
Electrical properties, 110, 117
Electrical resistance, 117
Electric arc furnace, 60-62
Electric induction furnace, 75-77
Element, 11, 12
Elements, periodic table of the, 17
Endurance strength, 112, 113
Etching, 161-163
Eutectoid point, 149

F

Face centered cubic space lattice, 124
Fatigue strength, 112, 113
Ferrite,
 appearance, 156
 structure, 124, 126, 131, 140
 transformation, 145, 146, 151
Ferrite-pearlite structure, appearance,
 156, 157
Ferritic iron, 126
File hardness testing method, 101, 102
Flame hardening, 233, 234
Flexure strength, 113
Flotation cells, 46
Foundries, steel, 42, 43
Full annealing, 183
Furnace,
 basic oxygen, 57-60
 Bessemer Converter, 53
 blast, 41, 47, 51-53
 cast iron-making, 42
 electric arc, 60-62
 electric induction, 75-77
 open hearth, 53-57
 steel-making, 41, 42, 53

G

Gas carburizing, 227, 228
General quenching and tempering, 126
Grain, 129
Grains and crystals, 18
Grain size, 151

effect of, 130, 131
Gray cast iron, 34, 35

H

Hardening,
case, 8, 223-243
surface, 223-243
Hardness, 82-107, 151, 152,
202-211
Hardness testing methods,
Brinell, 84-88
comparison, 102
conversion scales, 84, 85,
102-107
file, 101, 102
Knoop, 91, 92
Moh, 101
Rockwell, 92-97
Rockwell Superficial, 97, 98
Shore Scleroscope, 99, 100
Sonodur, 100
Vickers, 88-91
Hardness, units of, 84
Heat treating, 168-188, 213-243
Heat-treating techniques,
comparison, 221
Hematite, 46
High alloy steels, 32
High carbon steel, 26
Hypereutectoid region, 149
Hypoeutectoid region, 149

I

Impact strength, 113
Induction hardening, 234-237
Industrial I-T diagrams, 192, 195, 199-204
Industry, examples of metallurgy in, 8, 9
Ingot molds, 57
Ingot processing, 63-69
billets, 65
blooms, 65
rolling mill, 65, 66
slabs, 65
soaking, 64
stripping, 63, 64

teeming, 63
Internal stresses, 8, 9, 213-221
cracking, 137, 138, 213, 220
distortion, 8, 9, 217, 220
Iron, 23, 48-53
cast, 33-38, 72-76, 141
wrought, 23, 39
Iron-carbon phase diagram, 140-152, 189
Iron manufacture, 41-53
blast furnace, 47-52
pig iron, 41, 52, 53
Iron ore, 43-47
deposits, 43
mining, 43, 44
processes, 43-46
types, 46, 47
Iron, space lattice structures in, 126
Iron, wrought, 23, 39
Isothermal quenching and tempering,
219
Isothermal transformation diagrams,
189-211
Isothermal transformations, 193-199
I-T diagram,
basics, 189
C curves, 192
limitations, 190
making, 202, 203
purpose, 192, 193
regions of transformation, 193
temperature-time-line, 190
I-T diagrams,
comparison of different steels, 199
industrial, 192-195, 199-204

K

Knoop hardness testing method,
91, 92
Knoop units, 84

L

Ladles, 53
Lattice,
body centered cubic space, 123, 124
body centered tetragonal space, 125, 126